D0745805

visual **math**

visual math

See How Math Makes Sense

Jessika Sobanski

NEW YORK

Published in the United States by LearningExpress, LLC, New York.

Library of Congress Cataloging-in-Publication Data:
Sobanski, Jessika.
Visual math : see how math makes sense / by Jessika Sobanski.—1st ed.
p. cm.
ISBN 1-57685-404-3
1. Mathematics—Study and teaching. 2. Visual learning. I. Title.

QA11.2. S63 2002
510-dc21 2001050621

Printed in the United States of America
9 8 7 6 5 4 3 2 1
First Edition

ISBN 1-57685-404-3

For more information or to place an order, contact LearningExpress at:
900 Broadway
Suite 604
New York, NY 10003

Or visit us at:
www.learnatest.com

contents

visual **math**

Introduction

this book has been designed to allow learners to "see" how math makes sense. By combining logical math concepts with pictures, previously unclear images will fade and math will suddenly click for you. Read on to see how visual learning relates to the hemispheres of your brain and how combining images and logical reasoning actually gets both sides of your brain working at the same time, in the brain-healthy, whole brain learning style.

learning visually

When we look at the percentages of how much we learn through each sense, the breakdown looks like this:

- taste 3%
- smell 3%
- touch 6%
- sound 13%
- **sight 75%**

Although the following discussion of left brain versus right brain and whole brain learning strategies is fascinating, the premise of this book can basically be summed up by this old adage:

A picture is worth a thousand words!

left brain vs. right brain

Everybody has a left brain and a right brain, and we all use both sides. But most people use one side more than the other. This *hemispheric dominance* affects the way we process information and learn. Learning with both sides helps us make the most of our brains. Incorporating whole brain learning strategies into academic endeavors will address the *left-brainers* and *right-brainers* and allow both types to use more of their brains.

Would you rather look at this . . .

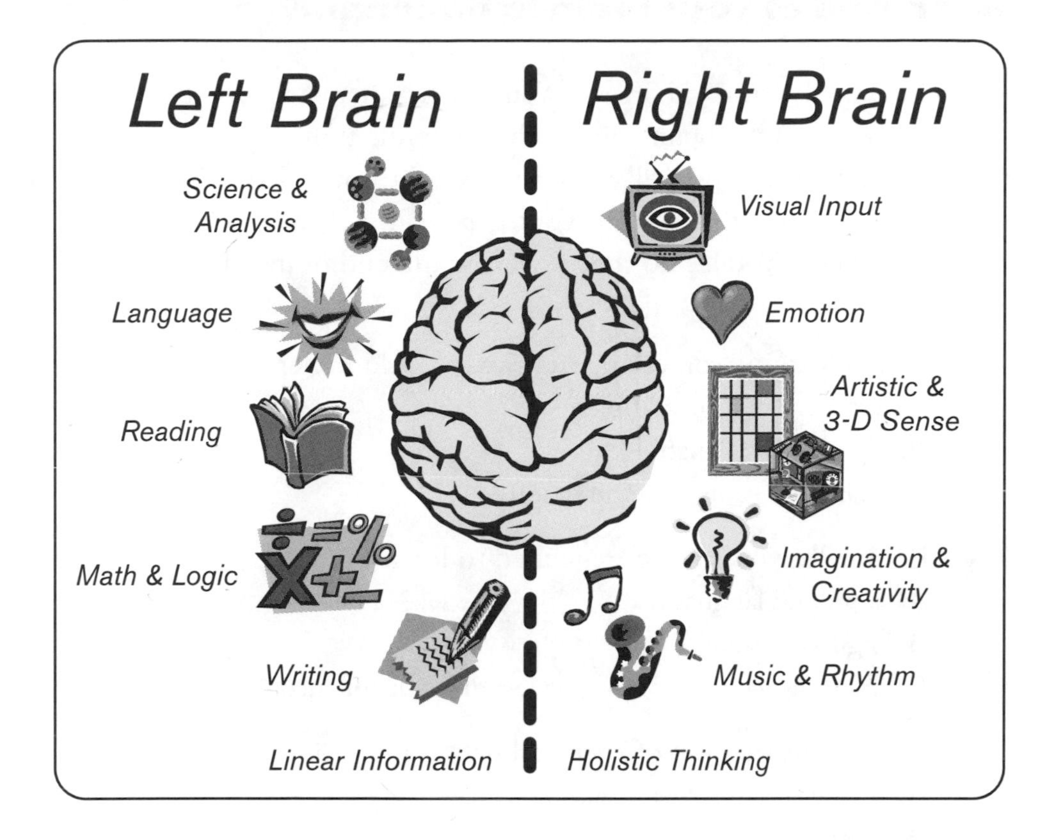

. . . or read this?

The processing in the left brain is linear. This means that learning occurs from part to whole. Processing in this hemisphere is also sequential. The left brain is good at processing symbols and is very logical and mathematical. The left brain also deals with verbal and written inputs and adheres to rules. Left brain processes are reality-based.

The processing in the right brain is holistic. This means that learning occurs by first envisioning the whole picture. Processing in this hemisphere is random. The right brain is also color-sensitive. This hemisphere is good at processing the concrete: things that can be seen, touched, and felt. The right brain is very intuitive and non-verbal. Right brain processes are fantasy-oriented.

which half of your brain is dominant?

The following questions are merely a survey that may help you better understand which side of the brain you emphasize while thinking, acting, learning, and so on. There are no wrong answers. Read through the choices given to you and pick the one that best fits your personality. Remember, accuracy depends on honesty. You'll find the answers at the end of the chapter.

1. When learning something new, you would rather
  A. learn by demonstration.
  B. learn by explanation.
  C. learn by reading the directions.

2. Personally, you are more inclined to learn
  A. a second language.
  B. sign language.
  C. Neither would be more or less difficult for me.

3. Which presentation of statistical data is more understandable?
  A. visual data, such as a graph or chart
  B. numerical data
  C. Both ways are just as understandable to me.

4. Which courses did you/do you enjoy most in school?
  A. philosophy/creative writing
  B. mathematics/science
  C. I was not partial to any particular course.

5. When choosing a movie to watch, you are more likely to enjoy
  A. a non-fiction documentary.
  B. a realistic "whodunnit" mystery film.
  C. a science fiction horror film.

6. The ideal activity on your night off is
  A. hanging out with a few close friends.
  B. sitting at home enjoying your favorite hobby.
  C. going dancing and meeting new people.

7. If choosing a vacation, you would
   A. choose a place you've never been to.
   B. choose the same place you went last year.
   C. choose a place similar to one you've gone to before.

8. Outside of special occasions, which best describes your wardrobe?
   A. relaxed, with your own personal sense of style
   B. neat and similar to that of others
   C. not interested in what other people think

9. When meeting new people, which personality trait most appeals to you?
   A. humor
   B. modesty
   C. intelligence

10. When planning a recreational activity, you would rather
    A. make long-term plans
    B. spontaneous plans
    C. It doesn't bother me either way.

11. When debating a subject you are passionate about, you
    A. let your emotions control the conversation.
    B. keep cool and collected, controlling your emotions.
    C. don't let your emotions play a factor.

12. When faced with a difficult decision, you
    A. make a decision influenced by a similar experience.
    B. make a decision based on instinct.
    C. find out all the info and make the best decision.

13. When it comes to workspace, which best describes you?
    A. a completely cluttered mess
    B. slightly messy but generally organized
    C. neat and organized

14. When engaged in a conversation, you tend to interpret participants' responses
    A. purely by the words they are saying.
    B. by body language only.
    C. both factors combined.

15. Immersed in thought while lying in bed ready to go to sleep, you are more likely to

A. think about what you want to dream about.
B. analyze the day's events.
C. plan ahead for tomorrow.

16. While driving home from a new job you realize there may be other routes to take. Which best describes you?

A. you'd stray from the usual path to find the most convenient route
B. you'd consult a map and take a new route the next day
C. you'd stay on the familiar route

17. If your boss at work gave you an unfamiliar task, would you

A. get ideas from someone who is familiar with the task and improve on them?
B. ask someone how best to get it done and follow his/her instructions?
C. develop your own technique?

18. When you go to a museum, which exhibits interest you the most?

A. artistic exhibits (paintings, sculptures)
B. antique exhibits (architecture, armor, relics)
C. prehistoric exhibits (extinct animals, prehistoric man, dinosaurs)

19. At a job interview, you would prefer the interviewer to ask

A. questions open for discussion.
B. questions requiring short specific answers.
C. questions that have short answers but allow you to add detail and substance.

20. Which adjective best describes you?

A. focused
B. independent
C. social
D. spacey

After completing the survey consult the **Brain Dominance Survey Key** (located on page 17 at the end of this chapter) to discover your brain dominance.

whole-brain learning strategies

No matter which hemisphere of your brain is dominant, keeping both hemispheres actively involved in the learning process will help you make the most of your brain. Here are some tips on creating a whole brain learning environment for yourself:

- Learn in a relaxed environment. The best recall occurs when brain wave patterns show a relaxed state.
- Learn in a multi-sensory environment by involving visual, auditory, and kinesthetic activities.
- Use color! This stimulates the right brain and helps recall.
- Make sure you take breaks every hour.
- Try to relate what you are learning to a bigger picture.
- Reinforce what you have learned through practice and review.

The following section contains "Brain Games" that are really good for your brain. Solving these puzzles requires the use of your whole brain. The concept is that you are using the logical (left) and visual (right) portions of your brain at the same time. You can solve a few now, and come back and try some more at a later time. Many standardized tests include questions that are based on logic, patterns, and sequences. These puzzles will help you foster those skills. In addition, as you read through this book, you will see that the logical concepts being presented are demonstrated in a visual manner. In other words, this book is good for your whole brain!

brain games

1. Bill, the neighborhood ice cream truck driver, is making his daily rounds. Today is extremely hot and muggy, and the kids around town are anxious for Bill to bring them ice cream. There is one problem: The hot and muggy weather has caused most of Bill's freezers to break down, leaving him with room for only 20 ice cream bars. Since Bill's house is just around the corner, he can always restock with more. The children usually wait in

the same spot for Bill to come around, but he does not want to pass them without any ice cream because he fears lost sales. Help Bill plan the best route to serve all the children and restock when he needs to without passing them by. Use the least number of trips because those kids are hot!

A number in a circle represents each group of kids. Each child will buy one ice cream bar and then leave. The arrows indicate one-way streets.

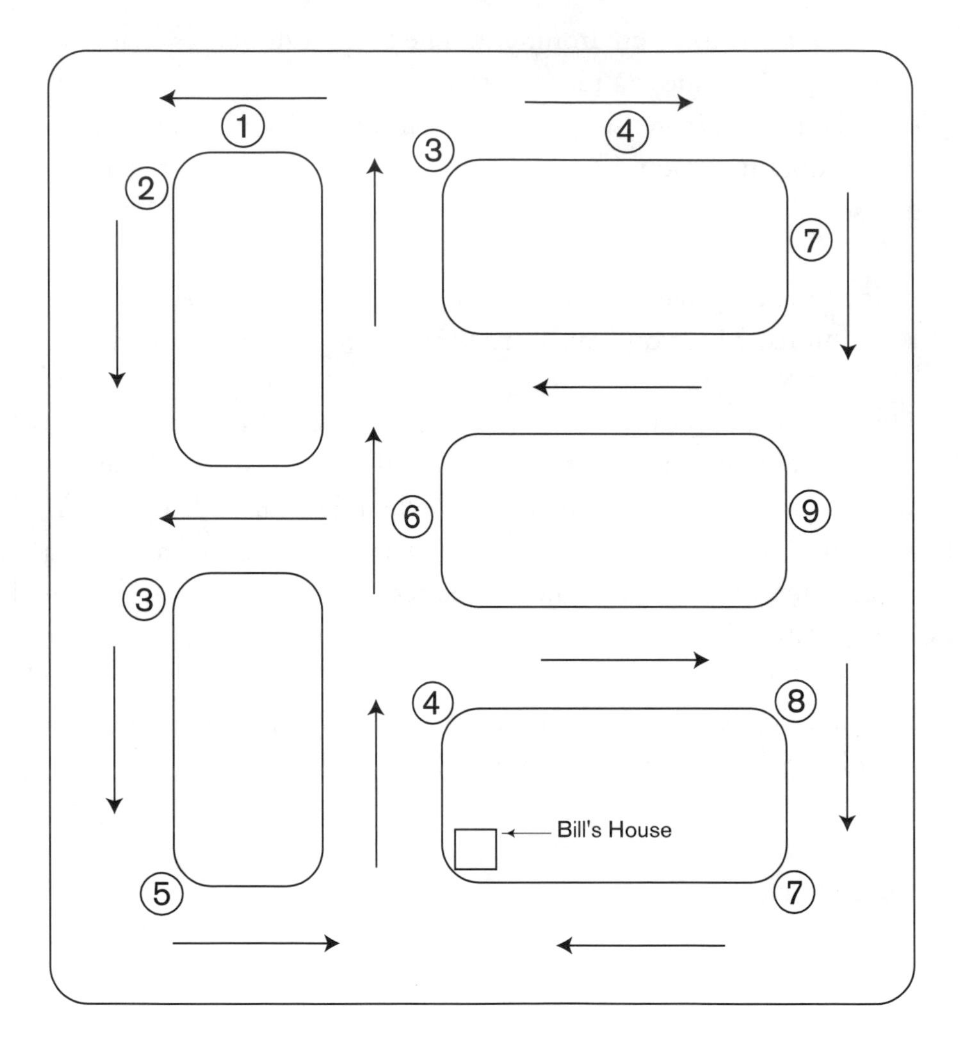

2. Can you place each labeled piece of the puzzle in the correct position?

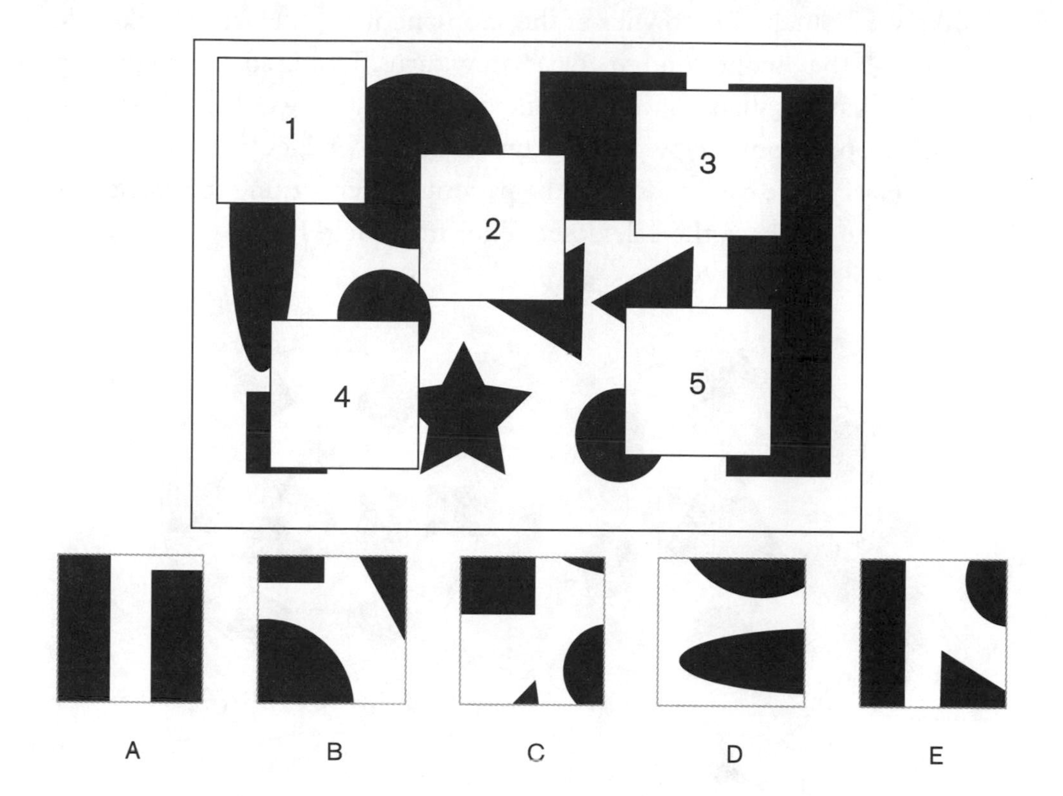

3. Little Harry finds his way to a curiosity shop and finds interest in a few knick-knacks. Among them is an old kaleidoscope, which Harry seems to like very much. The owner of the shop encourages Harry to take a look through the "scope" and to give it a few turns. Doing so, Harry sees a pattern of various shapes and colors and decides that he would like to purchase the "scope" from the owner. The owner acknowledges Harry's interest and tells Harry if he can guess what the pattern is after turning the kaleidoscope twice, he can have the kaleidoscope for free. Help Harry.

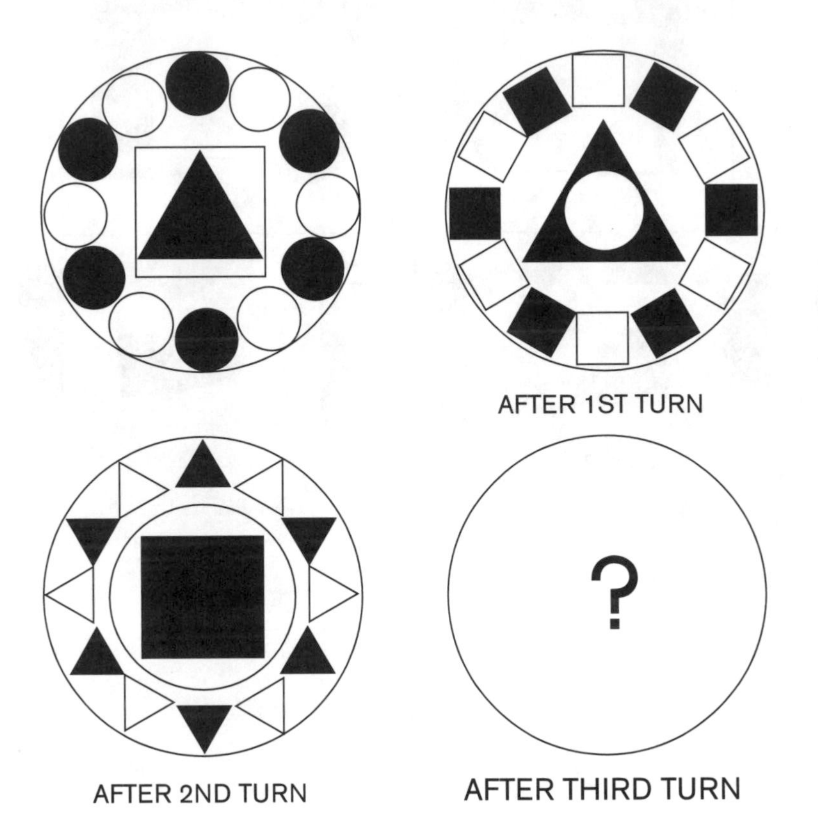

Judging from the choices given and the patterns in the kaleidoscope before and after each turn, pick the pattern that is most likely to appear when Harry turns the kaleidoscope again.

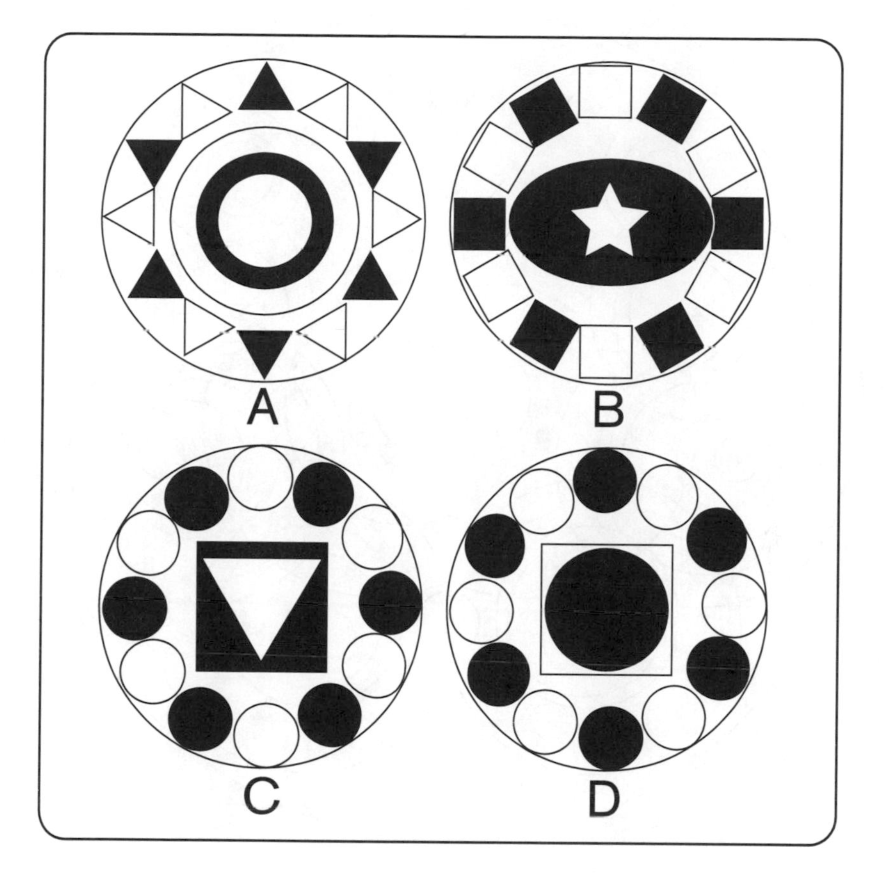

4. Margaret was working on an art project for school. The right side of the dashed line in the figure below should be symmetrical to the left side. However, 5 circles are not symmetrical. Can you find them?

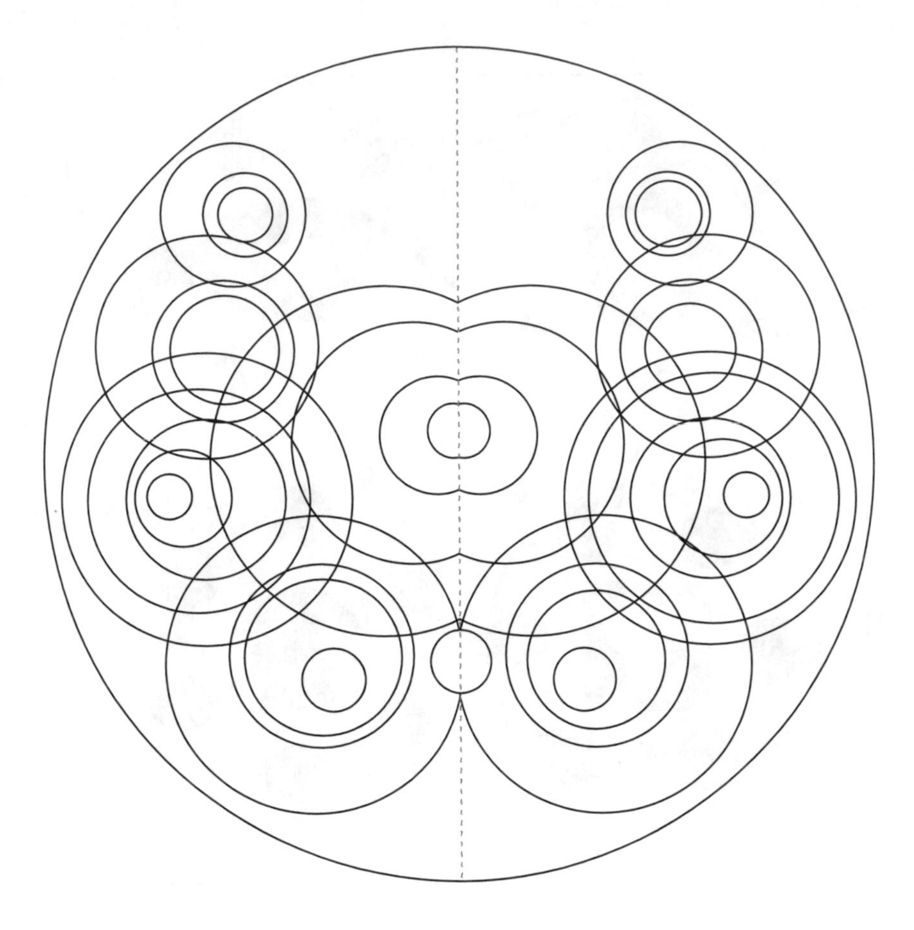

5. In the math puzzle below, there is a specific relationship between the numbers in the squares. Can you figure out the pattern and fill in the missing piece?

18	6	2	2
6	3	3	1
16	2	1	3
32	8	2	$\frac{1}{3}$
9	?	4	6

solutions to chapter exercises

1.

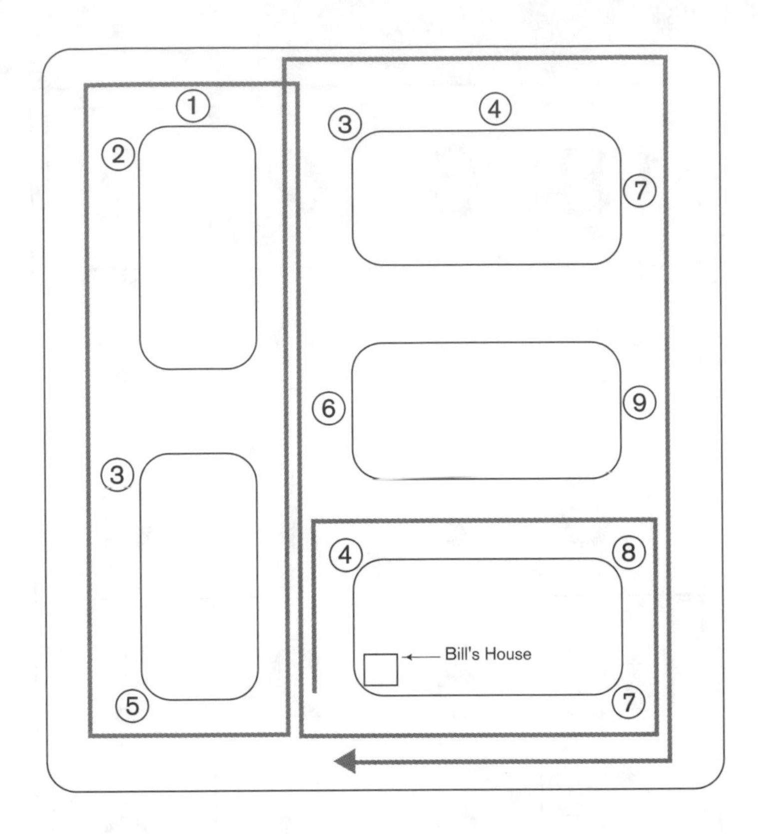

2.

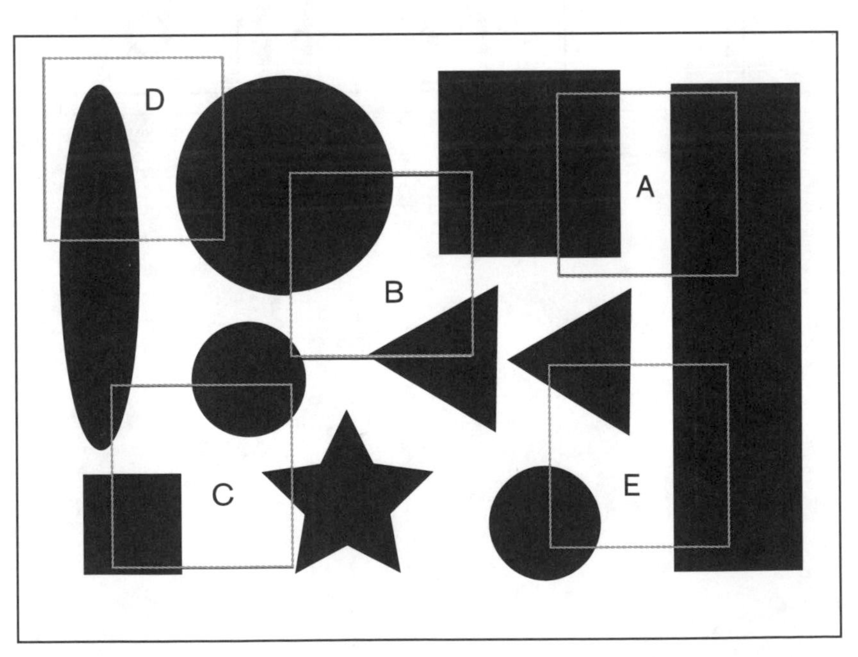

3. The correct answer is **c.** Notice the central shapes:

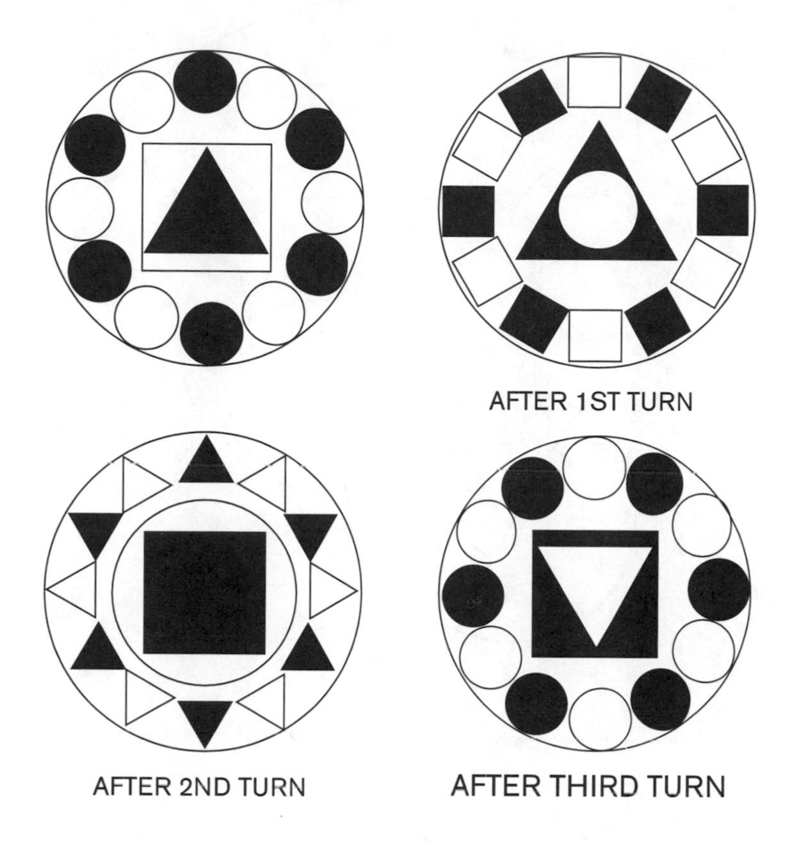

Also, if you notice the shape in the "12 o'clock" position of each figure, it goes from **black** circle **→** **white** square **→** **black** triangle, and thus you might expect the next "12 o'clock" shape to be **white**. A white circle would also make sense, as the pattern would go from circles in the periphery to squares, to triangles.

4.

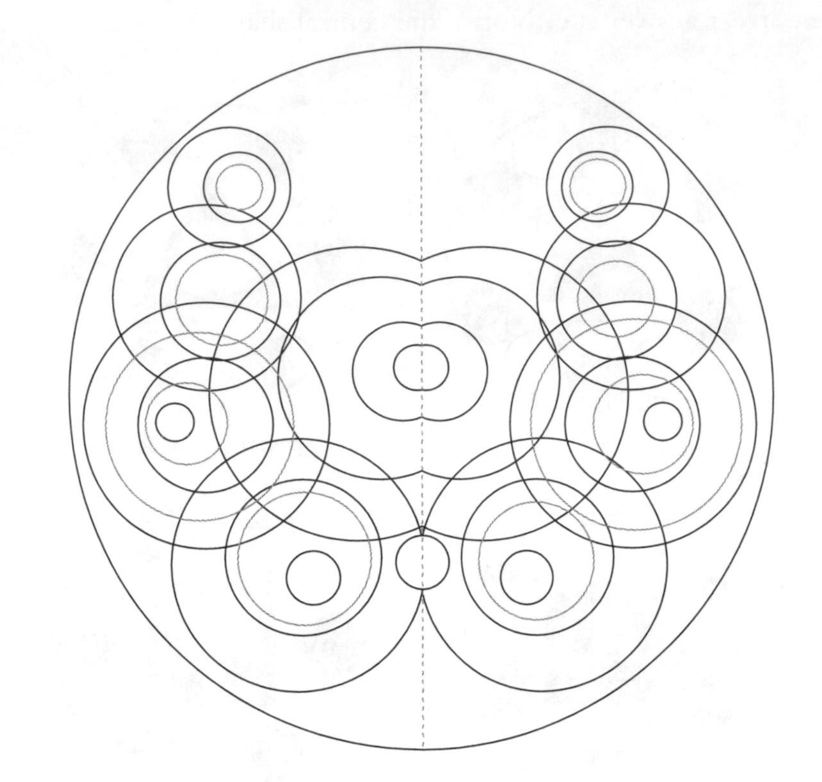

5.

18 ÷	6 is	2	2
6	3	3	1
16	2	1 ÷	3 is
32 ÷	8 is	2	$\frac{1}{3}$
9	?	4	6

brain dominance survey key

Add up your score from the Brain Dominance Survey on pages 4–6 based on the following point values for each answer choice

1. **A.** 4
B. 1
C. 2

2. **A.** 1
B. 5
C. 3

3. **A.** 4
B. 1
C. 3

4. **A.** 5
B. 1
C. 3

5. **A.** 5
B. 1
C. 3

6. **A.** 2
B. 1
C. 5

7. **A.** 5
B. 2
C. 4

8. **A.** 4
B. 2
C. 5

9. **A.** 5
B. 3
C. 1

10. **A.** 1
B. 5
C. 3

11. **A.** 5
B. 3
C. 1

12. **A.** 4
B. 5
C. 1

13. **A.** 4
B. 5
C. 1

14. **A.** 1
B. 5
C. 3

15. **A.** 5
B. 2
C. 1

16. **A.** 4
B. 2
C. 1

17. **A.** 1
B. 2
C. 5

18. **A.** 5
B. 3
C. 1

19. A. 5
B. 2
C. 4

20. A. 1
B. 1
C. 5
D. 5

If your score is between the numbers 23–32, your brain functions predominantly on the left side.

If your score is between the numbers 88–97, your brain functions predominantly on the right side.

If your score is between the numbers 33–54, your brain functions mostly on the left side but you have adapted, in some ways, to include some right brain attributes.

If your score is between the numbers 66–87, your brain functions mostly on the right side, but you have adapted, in some ways, to include some left brain attributes.

If your score is between the numbers 55–65, your brain functions on a bilateral level with equal right and left side attributes.

Number Concepts and Properties

getting number savvy

Numbers, numbers, numbers. Before we begin to explore mathematical concepts and properties, let's discuss number terminology. The counting numbers: 0, 1, 2, 3, 4, 5, 6, and so on, are also known as the **whole numbers**. No fractions or decimals are allowed in the world of whole numbers. *What a wonderful world*, you say. No pesky fractions and bothersome decimals.

But, as we leave the tranquil world of whole numbers and enter into the realm of **integers**, we are still free of fractions and decimals, but are subjected to the negative counterparts of all those whole numbers that we hold so dear. The set of integers would be:

$$\ldots . -3, -2, -1, 0, 1, 2, 3 \ldots .$$

The **real numbers** include any number that you can think of that is not imaginary. You may have seen the imaginary number ***i***, or maybe you haven't. The point is, you don't have to worry about it. Just know that imaginary numbers are not allowed in the set of real numbers. No pink elephants either! Numbers that are included in the real numbers are fractions, decimals, zero, π, negatives, and positives.

A special subset of the integers is the **prime numbers**. A prime number has just two positive factors: *one* and *itself*. It saddens many to realize that 1 is *not* prime by definition. Examples of prime numbers include: 2, 3, 5, 7, 11, 13, 17, 19, 23, and 29. Note that the opposite (negative version) of the above numbers are also prime. For example, the factors of −23 are 1, −23, −1, and 23. Thus, −23 is prime because it has exactly two positive factors: 1 and 23.

Not nearly as popular as the prime numbers are the **composite numbers**. Composite numbers have more than two factors. Note that 1 isn't composite either.

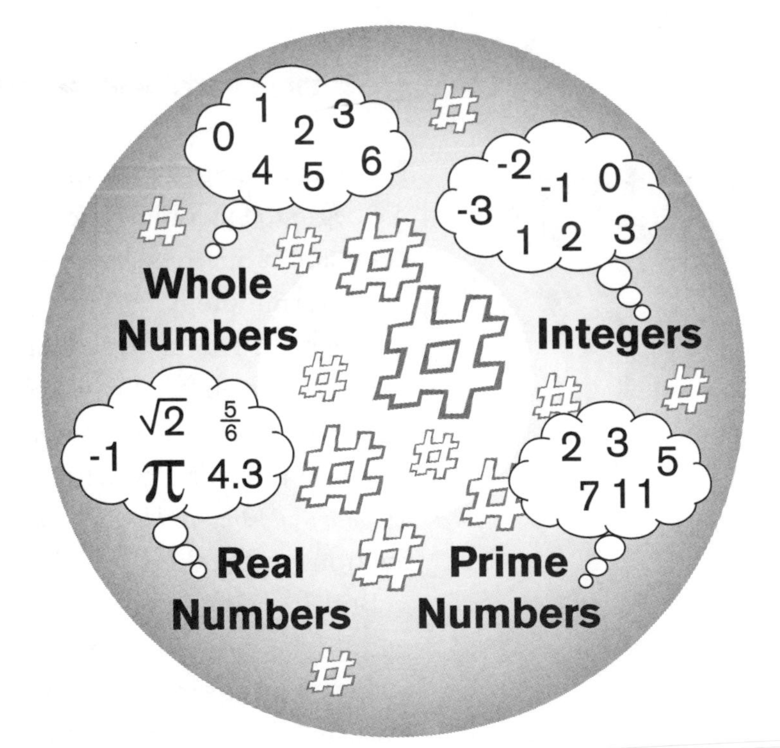

examples of different types of numbers

So where do irrational and rational numbers fit into all this? Here's how it works. The first five letters of RATIONAL are R-A-T-I-O. **Rational numbers** can be represented as a *ratio* of two integers. In other words, it can be written as a decimal that either *ends* or *repeats*. **Irrational numbers** can't be represented as a ratio, because their decimal extensions go on and on forever without repeating. π is the famous irrational number. Other irrational numbers are $\sqrt{2}$ and $\sqrt{11}$.

Your turn!

When you finish, you can find the answers at the end of the chapter, starting on page 48.

Exercise 1: Fill in the diagram below using the numbers 2, π, $2\frac{1}{2}$, 17, $-\frac{1}{3}$, .675, 79, 6, 1, −13, −555, and 8,700.

Note that the real numbers encompass all of the integers and whole numbers. Thus, if you place a number into the gray area labeled "whole numbers" you are also categorizing it as a real number and an integer.

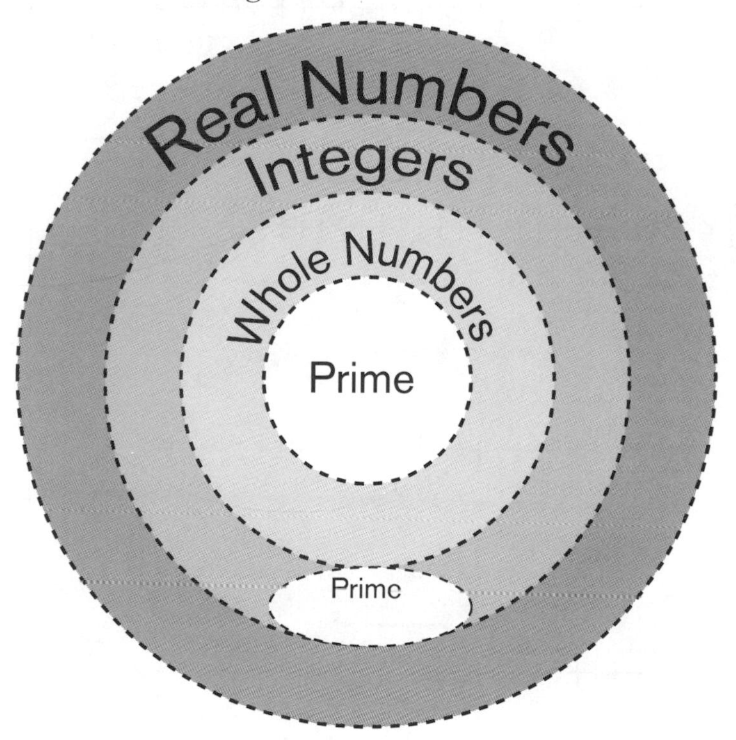

dealing with negatives

When working with negative values, it is helpful to think about a number line, a thermometer, or money. The following visual depicts addition and subtraction from different points on the number line.

addition and subtraction

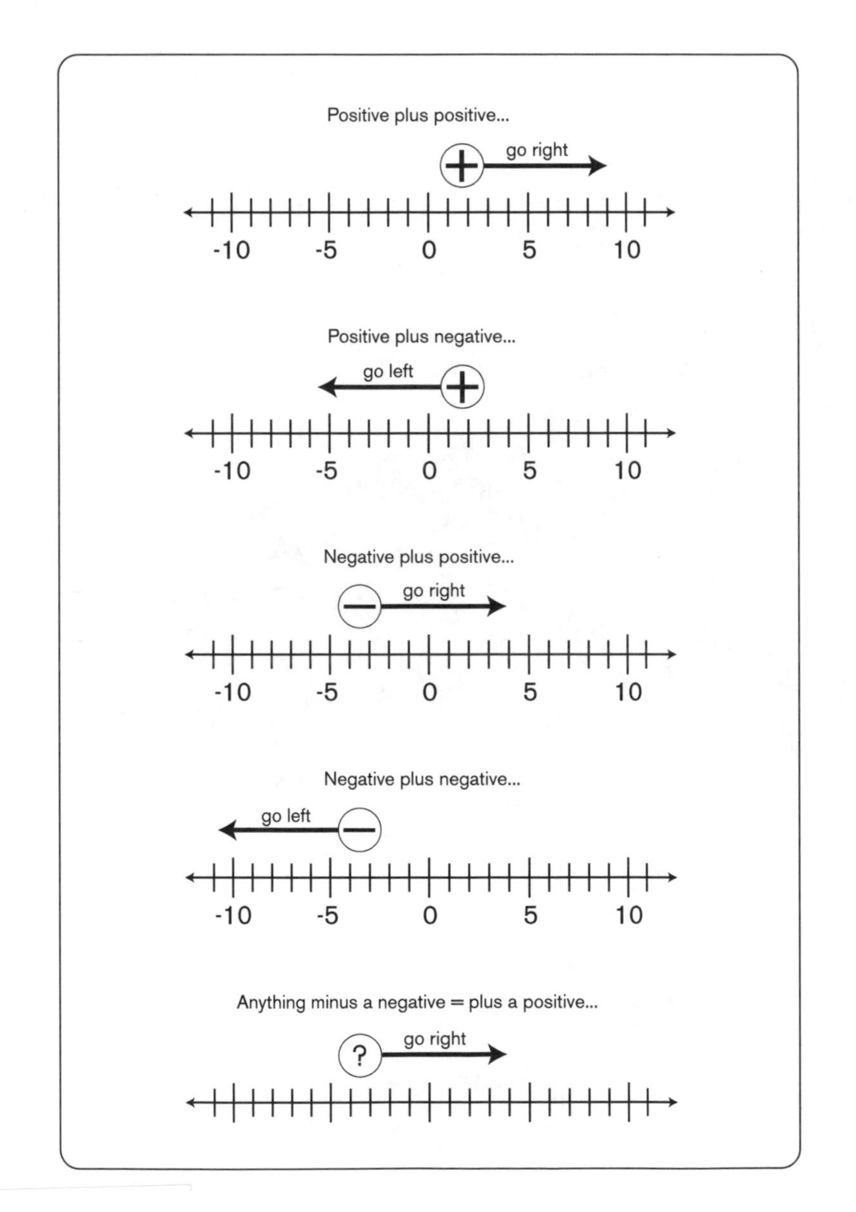

When multiplying and dividing signed numbers, you should be familiar with the rules below, where (+) is a positive integer, and (-) is a negative integer.

multiplication and division

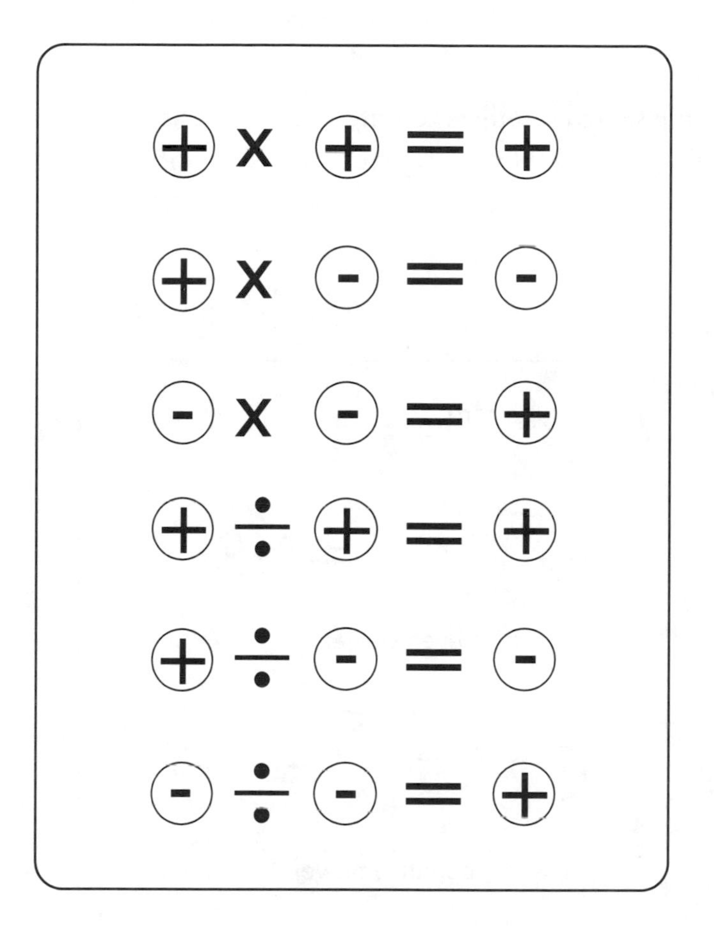

Another operation to consider for signed numbers is the effect of raising these numbers to different *powers*. The number in concern is the *base* and the power to which it is raised is the *exponent*. For example, when looking at 4^5, we call 4 the ***base*** and 5 the ***exponent***.

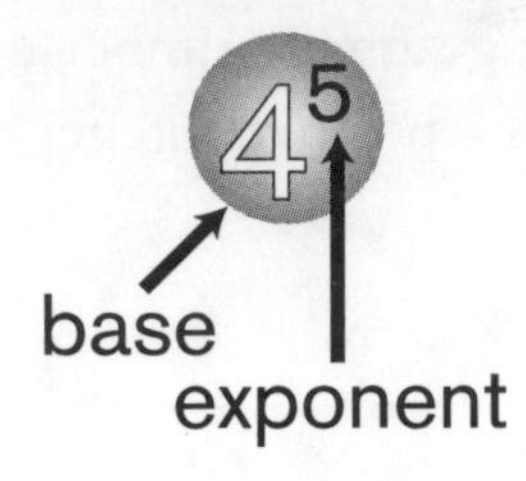

$$4^5 = 4 \times 4 \times 4 \times 4 \times 4$$

When raising signed numbers to different powers, it is important to bear the following rules in mind. Here, (+/-) represents a positive or negative number.

exponents

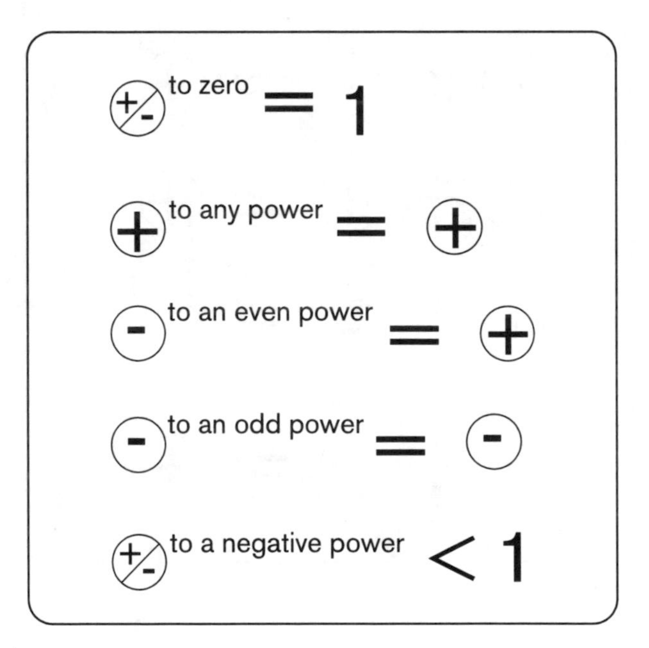

Some examples:

$5^0 = 1$

$7^2 = 7 \times 7 = 49$

$(-8)^2 = 64$

$(-3)^3 = -3 \times -3 \times -3 = 9 \times -3 = -27$

When dealing with negative exponents, remember that $a^{-n} = \frac{1}{a^n}$. For example:

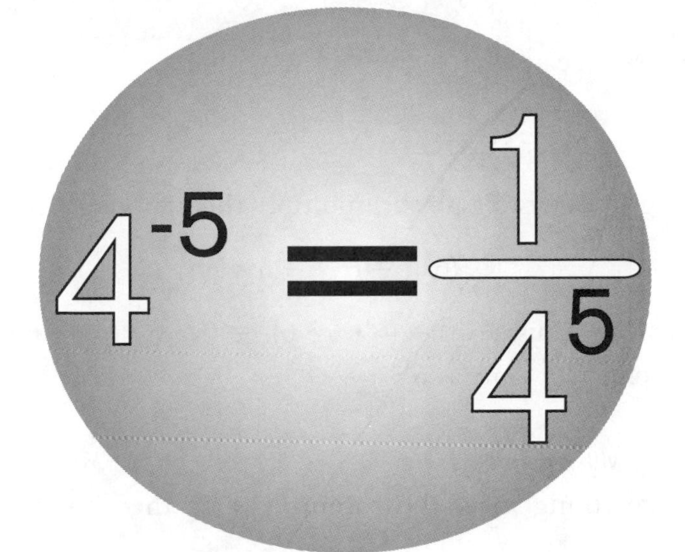

Some examples:

$4^{-2} = \frac{1}{4^2} = \frac{1}{16}$

$-2^{-3} = \frac{1}{-2^3} = \frac{1}{-8} = -\frac{1}{8}$

Rules for operations with exponents:

When multiplying powers of the same base, add the exponents.

$$4^2 \bullet 4^3 = 4^{2+3} = 4^5$$

Notice that this rule works because $4^2 \bullet 4^3 = (4 \bullet 4) \bullet (4 \bullet 4 \bullet 4)$, which is the same as 4^5.

When dividing powers of the same base, subtract the exponents.

$$5^8 \div 5^2 = 5^{8-2} = 5^6$$

This rule works because $\frac{5^8}{5^2} = \frac{5 \bullet 5 \bullet 5 \bullet 5 \bullet 5 \bullet 5 \bullet 5 \bullet 5}{5 \bullet 5}$

We can cancel: $\frac{\cancel{5} \bullet \cancel{5} \bullet 5 \bullet 5 \bullet 5 \bullet 5 \bullet 5 \bullet 5}{\cancel{5} \bullet \cancel{5}} = 5^6$

When raising a power to a power, multiply the exponents.

$$(6^3)^4 = 6^{3 \bullet 4} = 6^{12}$$

This rule works because $(6^3)^4 = (6 \bullet 6 \bullet 6)^4 = (6 \bullet 6 \bullet 6)(6 \bullet 6 \bullet 6)(6 \bullet 6 \bullet 6)(6 \bullet 6 \bullet 6)$, which is the same as 6^{12}.

Tips when dealing with exponents:

Always raise everything *inside* the parentheses by the power *outside* of the parentheses.

$$(7 \bullet 6^3)^4 = 7^4 \bullet 6^{3 \bullet 4} = 7^4 \bullet 6^{12}$$

$$(5a^3)^2 = 5^2 \bullet a^{3 \bullet 2} = 25a^6$$

When you do not have the same base, try to convert to the same base:

$$25^4 \bullet 5^{12} = (5^2)^4 \bullet 5^{12} = 5^8 \bullet 5^{12} = 5^{20}$$

Exercise 2: Simplify the following.

$3^2 \bullet 3^4 \bullet 3 =$ ____

$\frac{7^{10}}{7^5} =$ ____

$(2^3b^7)^2 =$ ____

$5^2 \div 5^{-2} =$ ____

$4^2 \bullet 2^5 =$ ____

$4^{-2} \div 4^{-5} =$ ____

place value

Each place that a digit occupies within a number has a name. We typically see numbers in **base 10**. In base 10 there are ten possible digits: 0, 1, 2, 3, 4, 5, 6, 7, 8, 9. When you have more than 9, you need to add another spot, namely, the tens place. 10 represents no ones and 1 ten. Below are all the places:

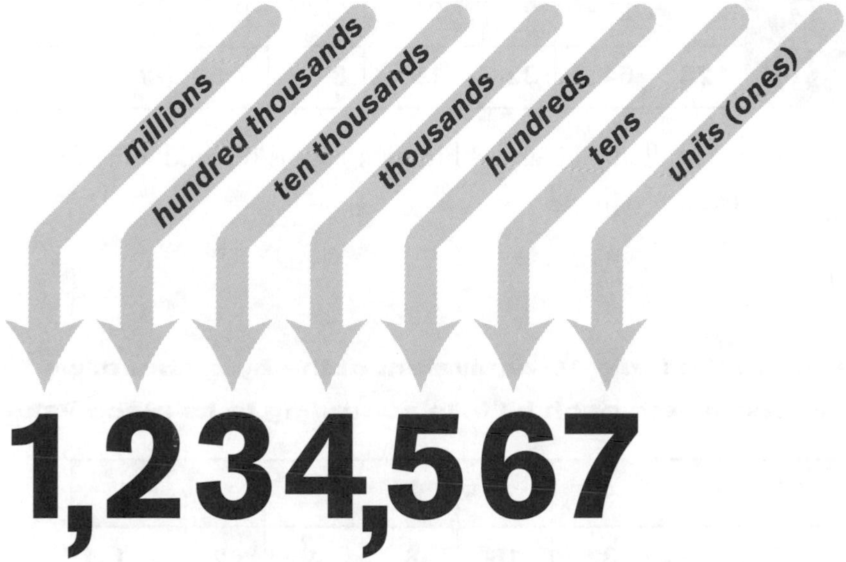

millions
hundred thousands
ten thousands
thousands
hundreds
tens
units (ones)

1,234,567

The number above can be represented in *expanded notation* as $(1 \times 1{,}000{,}000) + (2 \times 100{,}000) + (3 \times 10{,}000) + (4 \times 1{,}000) + (5 \times 100) + (6 \times 10) + (7 \times 1)$.

Note that in base 10, *ones* represent $10^0 = 1$, the *tens* represent $10^1 = 10$, the *hundreds* represent $10^2 = 100$, and so forth.

Base 10:

millions	hundred thousands	ten thousands	thousands	hundreds	tens	ones
10^6	10^5	10^4	10^3	10^2	10^1	10^0
1,000,000	**100,000**	**10,000**	**1000**	**100**	**10**	**1**

Base 2 uses only 2 digits: 0 and 1. If you've been around computers, you have probably heard the term *binary*. Even the most complex functions that a computer provides boils down to a data stream consisting of zeros and ones (base 2). The 1 stands for *on* and the 0 stands for *off*.

Here are the places in base two with the base ten equivalents noted underneath:

Base 2	2^7	2^6	2^5	2^4	2^3	2^2	2^1	2^0
Base 10	**128**	**64**	**32**	**16**	**8**	**4**	**2**	**1**

A solo 0 or a 1 is called a *bit* and 8 bits in a row is termed a *byte*. Consider the following *byte*:

10110101

To figure out the base 10 equivalent of the byte above, we first have to see where each bit falls according to its place value:

2^7	2^6	2^5	2^4	2^3	2^2	2^1	2^0
128	**64**	**32**	**16**	**8**	**4**	**2**	**1**
1	0	1	1	0	1	0	1

Next, we add up the values that are indicated:

2^7	2^6	2^5	2^4	2^3	2^2	2^1	2^0
128	**64**	**32**	**16**	**8**	**4**	**2**	**1**
0	1	1	0	0	1	0	1
128	+ 64	+ 32			+ 4		+ 1

128 + 64 + 32 + 4 + 1 = 229, so $\mathbf{10110101_{two}} = \mathbf{229_{ten}}$.

Exercise 3: Express 6,871,235 in expanded notation: ______________________

____

roots

What is the value of $4^{\frac{1}{2}}$? How about $27^{\frac{1}{3}}$? Looks complicated, huh? Those two questions could actually be rewritten as $\sqrt{4}$ and $\sqrt[3]{27}$. Let's take a closer look at roots and how to convert icky fractional exponents into pleasant little radicals.

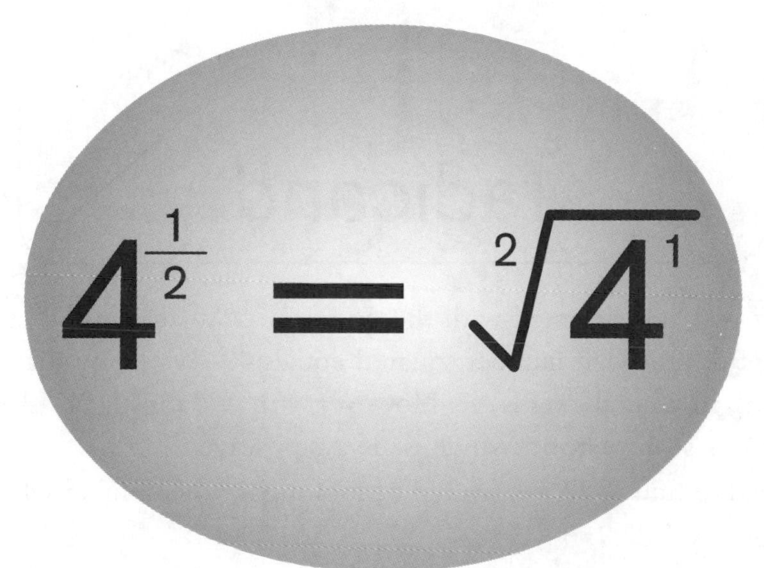

Some examples:

$25^{\frac{1}{2}} = \sqrt[2]{25}$

$8^{\frac{1}{3}} = \sqrt[3]{8}$

$16^{\frac{1}{2}} = \sqrt[2]{16}$

Note that $\sqrt[2]{}$ is the same as $\sqrt{}$. The radical symbol indicates the root to be taken (the **index**). If there is no index labeled, take the square root.

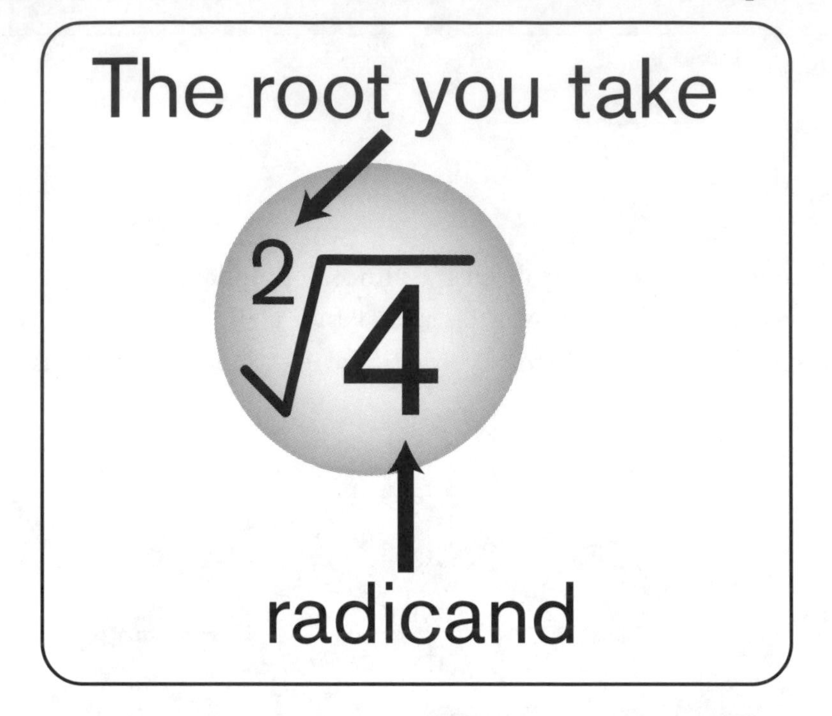

Let's say you are presented with the question: "What is $\sqrt{49}$?" Here, you need to figure out what number squared equals 49. We know that $-7^2 = 49$ and $7^2 = 49$, so your answer is ± 7. However, you will mostly be asked to find the **principal square root**, which is always positive.

If you had to find $\sqrt[3]{27}$, you would be looking for the number which, when cubed, would yield 27; $3 \times 3 \times 3 = 27$, so $\sqrt[3]{27} = 3$.

Exercise 4: Complete the chart below.

exponential form	rewrite as a radical	solve
$125^{\frac{1}{3}}$		
$121^{\frac{1}{2}}$		
$64^{\frac{1}{6}}$		
$-8^{\frac{1}{3}}$		

ways to manipulate radicals

1. You can express the number under the radical as the product of other numbers. You can then equate the root of the product of those numbers as the product of separate roots of those numbers.
   $\sqrt{12} = \sqrt{4 \cdot 3} = \sqrt{4} \cdot \sqrt{3} = 2\sqrt{3}$

2. When multiplying two roots (with the same index) you can combine them under the same radical.
   $\sqrt{8} \cdot \sqrt{2} = \sqrt{8 \cdot 2} = \sqrt{16} = 4$

3. When dividing two roots (with the same index) you can combine them under the same radical.
   $\sqrt{242} \div \sqrt{2} = \sqrt{242 \div 2} = \sqrt{121} = 11$

4. You can take a "division problem" out from under the radical and place each "piece" under its own radical.
   $\sqrt{\frac{15}{4}} = \sqrt{15} \div \sqrt{4} = \sqrt{15} \div 2$

5. You can only add roots if they have the same index and radicand.
   Recall:

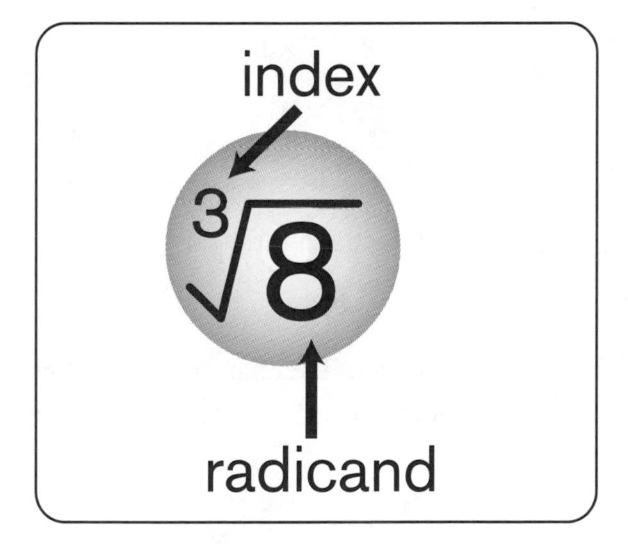

   $2\sqrt{3} + 3\sqrt{3} = 5\sqrt{3}$

6. Also, you can only subtract roots if they have the same index and radicand.
   $8\sqrt{5} - 2\sqrt{5} = 3\sqrt{5}$

order of operations

When you get a messy mathematical expression involving every operation under the sun, it is important to perform the operations in the correct order. The order of operations is:

1. parentheses
2. exponents
3. multiplication/division
4. addition/subtraction

Many people say, “Please excuse my dear Aunt Sally,” or “PEMDAS” in order to remember the correct order of operations.

Should we bother with order of operations? Does order matter?

If we took a problem like $6 + 8 \times 2 - 3 \times 5$ and just took each operation in order of appearance, we’d get:

$$\mathbf{6 + 8} \times 2 - 3 \times 5$$

$$\mathbf{14 \times 2} - 3 \times 5$$

$$\mathbf{28 - 5} \times 5$$

$$\mathbf{23 \times 5}$$

$$115$$

Is this right?

No.

Let’s proceed in the correct order. There are no parentheses or exponents, so we need to do any multiplication or division first (in the order in which they occur) . . .

$$6 + \mathbf{8 \times 2} - 3 \times 5$$

$$6 + 16 - 3 \times 5$$

We still have a multiplication to take care of . . .

$$6 + 16 - \mathbf{3 \times 5}$$

$$22 - \mathbf{15}$$

$$7$$

Notice that when you threw caution to the wind you got an answer that wasn't even close to the actual value! Take your time and carry out each operation in the correct order!

Let's go through a harder question step by step.

$\mathbf{5 + 2^2 \times 8 - (5 \times 3^2)}$	
Parentheses	$5 + 2^2 \times 8 - \mathbf{(5 \times 3^2)}$ When we look inside the parentheses, we first deal with the exponent rather than the multiplication because the E comes first in PEMDAS: $5 + 2^2 \times 8 - \mathbf{(5 \times 9)}$ Next, we multiply: $5 + 2^4 \times 8 - \mathbf{(45)}$
Exponents	$5 + \mathbf{2^2} \times 8 - 45$ $5 + 4 \times 8 - 45$
Multiplication/Division	$5 + \mathbf{4 \times 8} - 45$ $5 + \mathbf{32} - 45$
Addition/Subtraction	$\mathbf{5 + 32} - 45$ $\mathbf{37} - 45$ -8

Exercise 5: Use the chart below to sequentially follow the steps in PEMDAS.

$2^3 - 7 \times (5 - 8) \div 3$	
Parentheses	
Exponents	
Multiplication/Division	
Addition/Subtraction	

logarithms

Logarithms, or logs, can be to different bases. $\mathbf{log_2}$ denotes a log to the base 2, and $\mathbf{log_{10}}$ denotes a log to the base 10. Logarithms are exponents. When you solve a log, you are actually calculating the exponent that the base was raised to.

Look at this problem:

$$\log_2 4 = ?$$

Logs can be tackled easily by making a spiral right through the problem . . . .

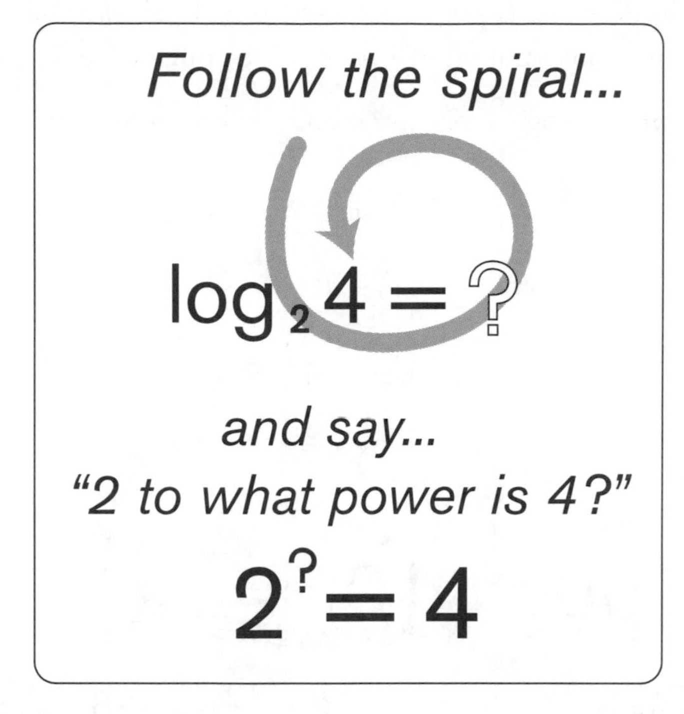

You spiral your way through the 2, then the "?" and end on the 4. "2 to what power is 4?" 2 to the second power is 4, so, $\log_2 4 = 2$.

$\log_{10}$ is so common that if you see a question like log 100 = ?, you are taking the log to the base 10. Thus $\log 100 = \log_{10} 100$. To solve, make a spiral . . . .

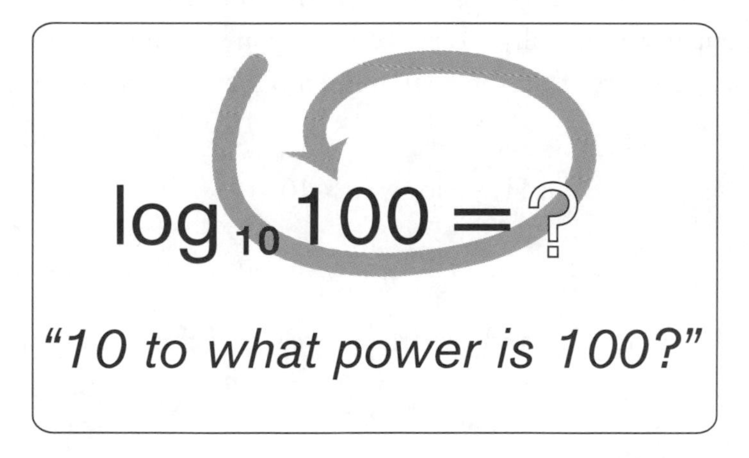

The number 10 to the second power is 100, so $\log_{10} 100 = 2$.

Are logs just fun with spirals? You ask. Logs are used in many branches of math and science. For example, in chemistry you study pH. Well, pH is just the negative log of the concentration of *hydronium ions*. (That's water with an extra positive charge added, or H_30+.) Let's say you had a *mystery solution* that had a hydronium ion concentration of 1×10^{-6}, we could solve for pH:

$$pH = -\log_{10} \text{(concentration of hydronium ions)}$$

$$pH = -\log_{10} (1 \times 10^{-6})$$

$$= -\log_{10} (10^{-6})$$

Let's calculate the log first . . .

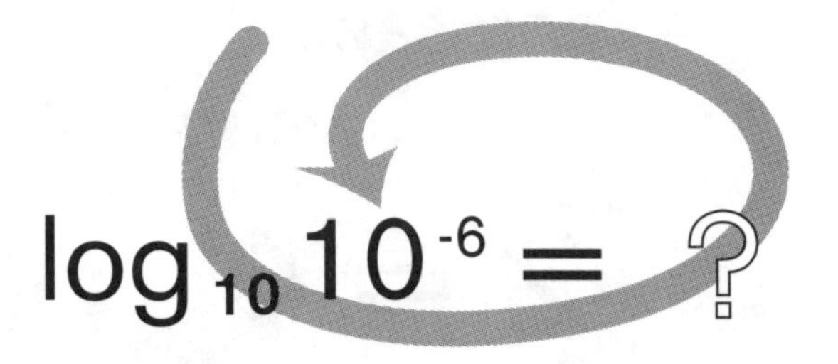

"10 to what power is 10^{-6}?"

Asking "10 to what power is 10^{-6}?" is kind of like asking, "Who's buried in Grant's tomb?" or "What color was George Washington's white horse?" Obviously, 10 to the $^{-}6^{th}$ power is 10^{-6}. See, this question looked ominous and menacing, but it was actually easier than any you've done so far! But before you start patting yourself on the back, remember to put the –6 back into the equation . . .

$$pH = -\log_{10} (1 \times 10^{-6}) =$$

$$pH = -(-6) = 6.$$

You could then look at a pH chart to reveal that this *mystery solution* is an acid!

Exercise 6:

Use a spiral to solve the logs below:

$\log_{10} 10{,}000 = ?$

$\log_{10} 1{,}000 = ?$

pre-algebra: a review of number properties

The **commutative property** holds for *addition* and *multiplication*, as shown below:

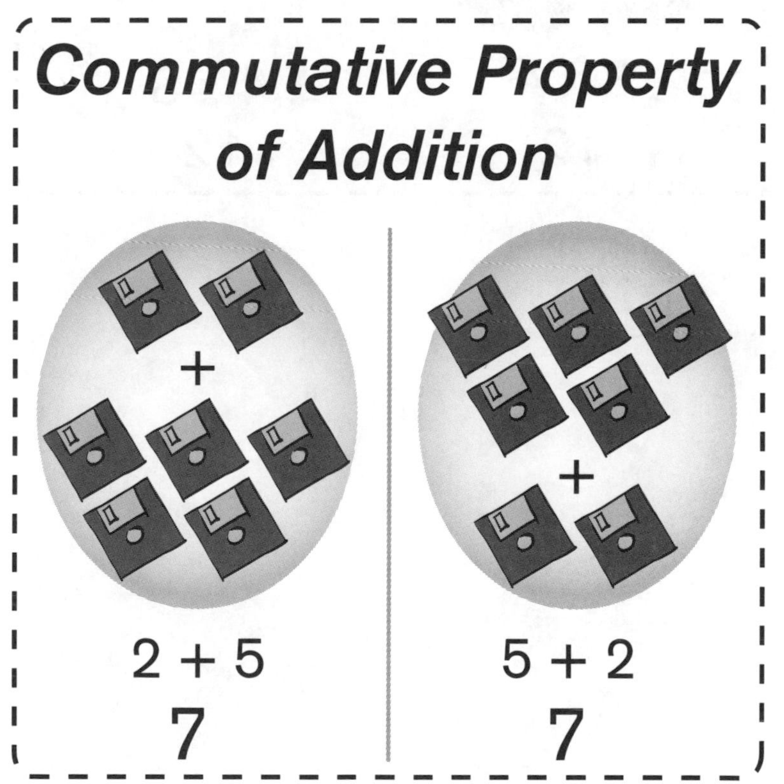

Note that $2 - 5$ does not equal $5 - 2$. If you want to apply the commutative property, you must rewrite $2 - 5$ as $2 + {}^{-}5$. Now the commutative property holds: $2 + {}^{-}5 = {}^{-}5 + 2$.

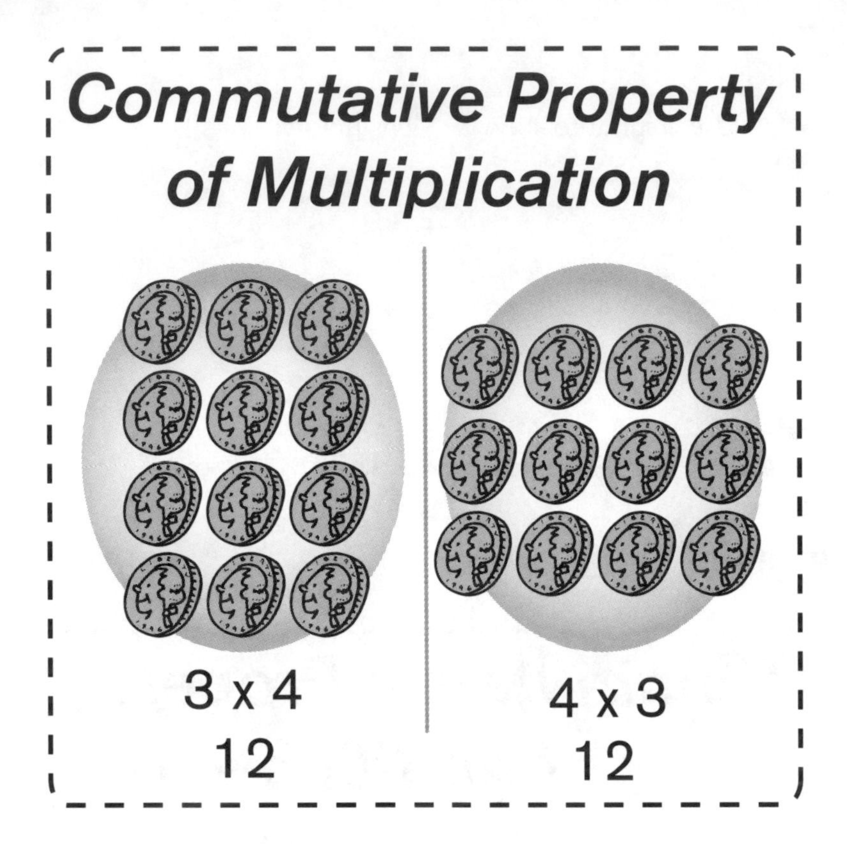

Another property that holds for addition and multiplication is the **associative property.**

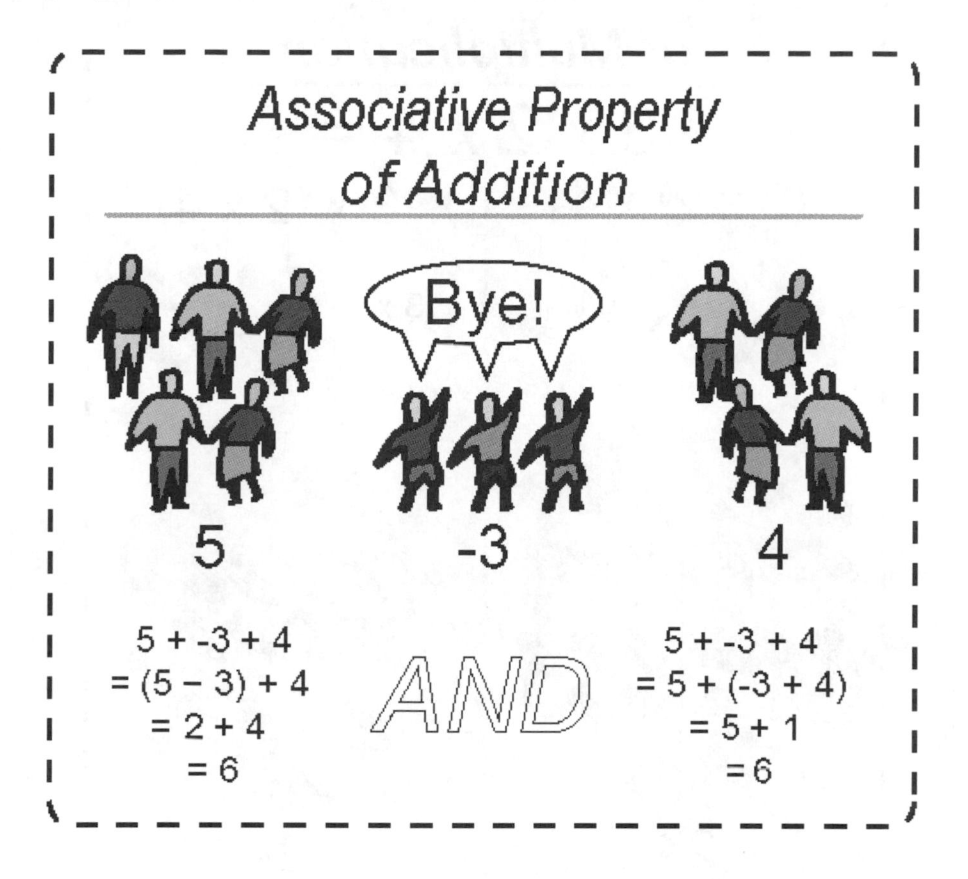

Note that $(5 - 3) + 4 = 6$, but $5 - (3 + 4) = 5 - 7$, or -2. The associative property holds for addition. In order for you to apply it to subtraction, you need to change $5 - 3 + 4$ into $5 + {}^{-}3 + 4$.

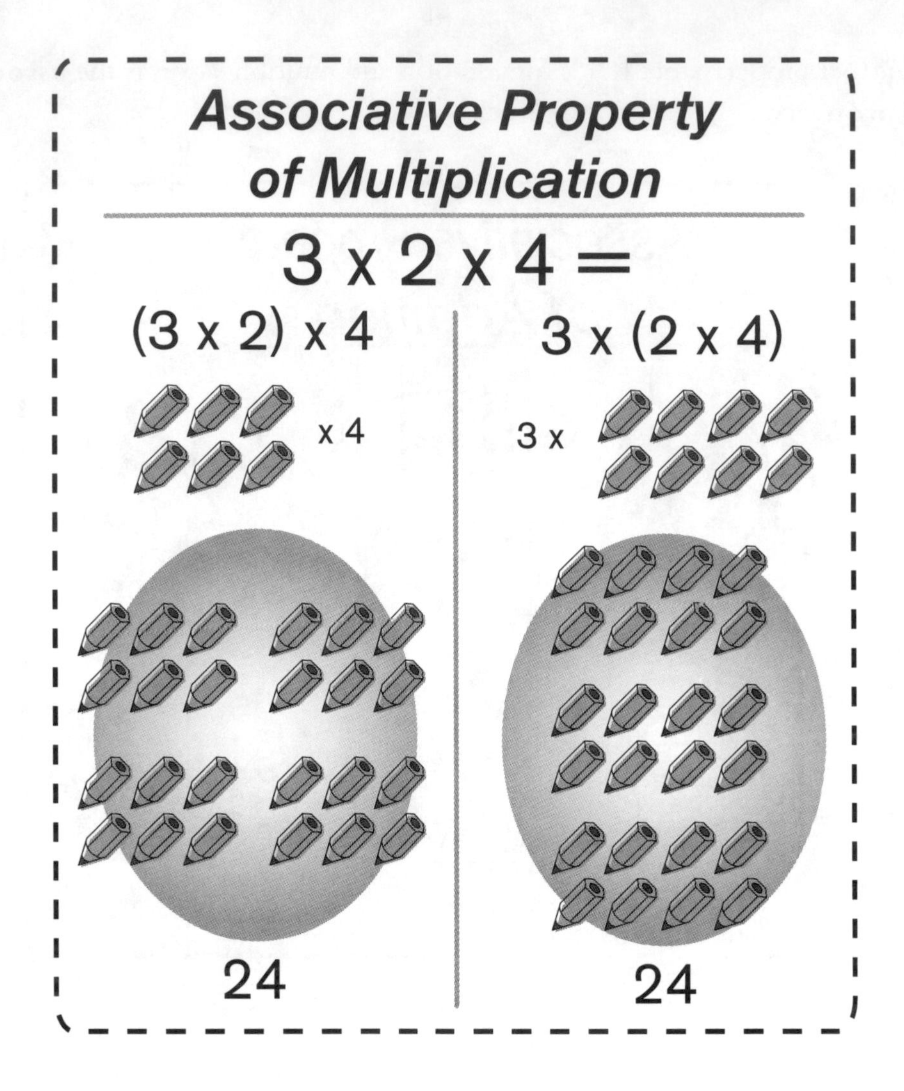
Associative Property
of Multiplication
3 x 2 x 4 =
(3 x 2) x 4
3 x (2 x 4)
x 4
3 x
24
24

When trying to calculate math problems in your head, the **distributive property** comes in handy. Suppose you wanted to give out awards and you lined up medals on a table in a 20 by 14 arrangement. How many medals do you have?

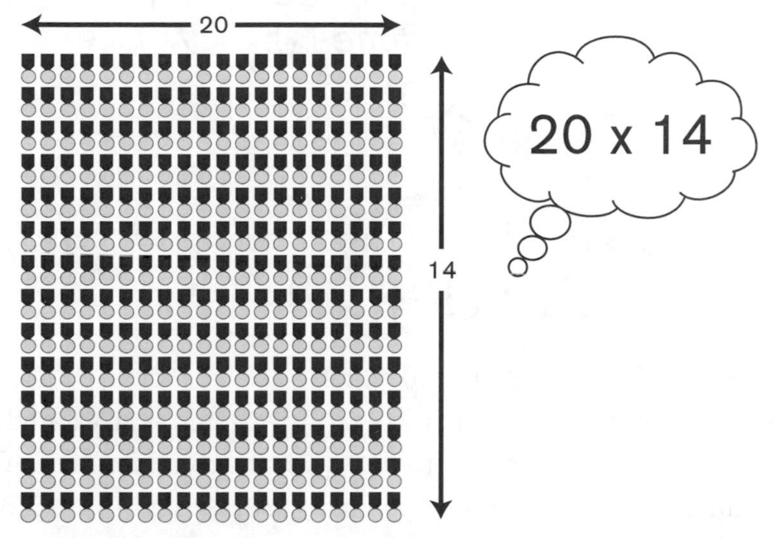

You can think of these medals as two groups . . .

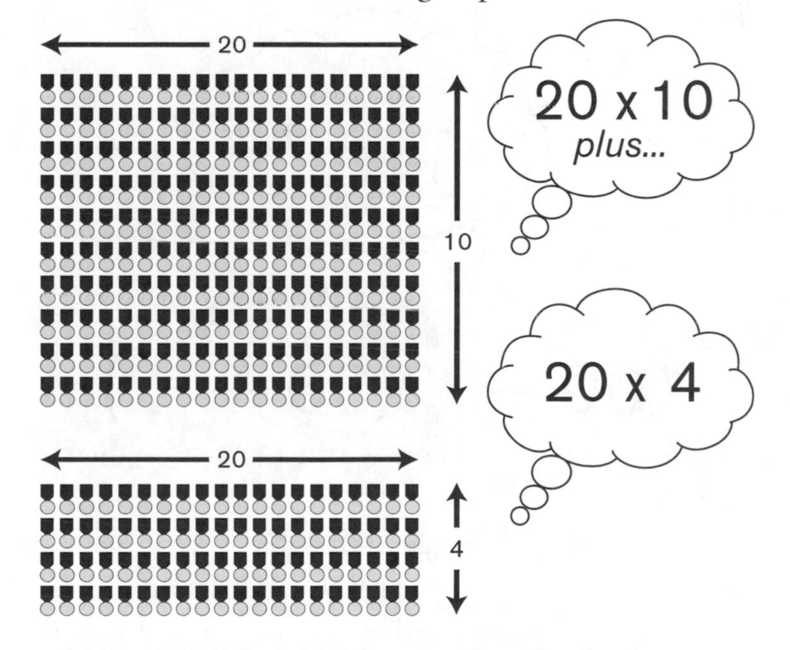

So, if you think of 20(14) as 20(10 + 4). The distributive property equates 20(10 + 4) with 20 • 10 + 20 • 4 . . .

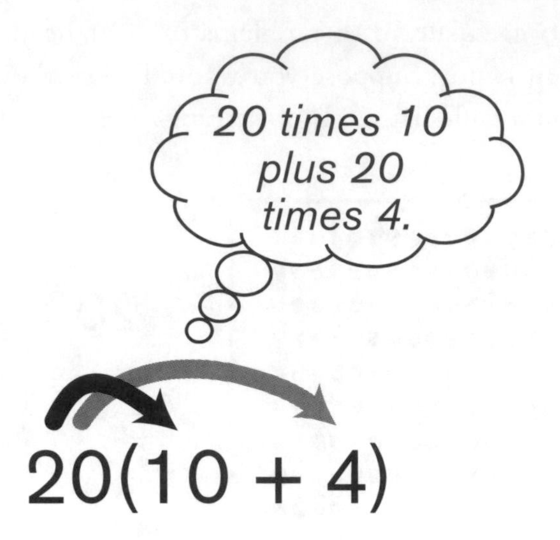

Thus, $20 \cdot 10 + 20 \cdot 4 = 200 + 80 = 280$.

Let's say you had the expression $100 - 2(30 - 2)$ and you wanted to apply the distributive property. One way to deal with the two instances of subtraction is to change each "*minus*" into "*plus a negative*." Thus, $100 - 2(30 - 2) = 100 + {}^-2(30 + {}^-2)$. Next, apply the distributive property . . .

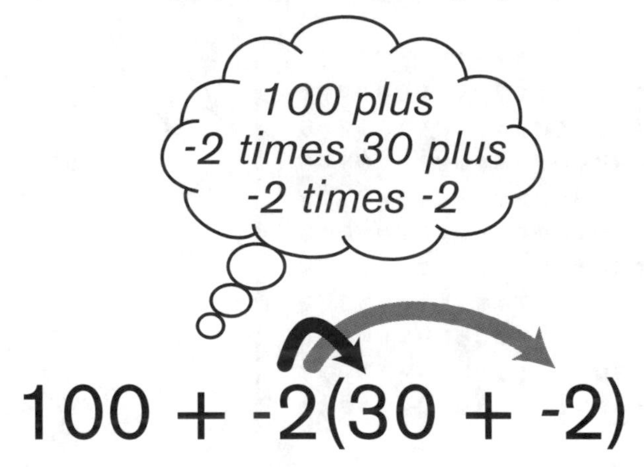

100 plus $^-2$ times 30 plus $^-2$ times $^-2$ equals $100 + ({}^-2 \cdot 30) + ({}^-2 \cdot {}^-2)$, or $100 - 60 + 4 = 44$.

You may have noticed that if given $100 - 2(30 - 2)$, it would be easier to just start inside the parentheses to yield $100 - 2(28)$. In this case, it's a viable option. But what if you had an *unknown*? Let's say you had $12 - (x - 1)$. First, you could get rid of the subtractions by turning each "*minus*" into "*plus a negative*." So, $12 - (x - 1) = 12 + {}^-1(x + {}^-1)$.

[Note that there was an "*invisible one*" in front of the parentheses $12 - (x - 1) = 12 - \mathbf{1}(x - 1) = 12 + {}^-1(x + {}^-1)$.]

Next, apply the distributive property . . .

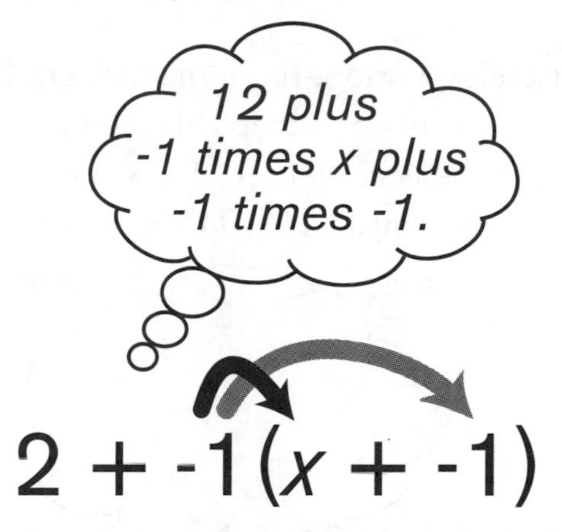

12 plus $^{-}1$ times x plus $^{-}1$ times $^{-}1$ equals $12 + {}^{-}1x + ({}^{-}1\ ({}^{-}1)$, or $12 - x + 1$.

Once you get used to turning a "*minus*" into "*plus a negative*" and spotting "*invisible ones*" you will be able to account for these things in your head. Thus, upon seeing $12 - (x - 1)$, you will think:

12 minus x
plus 1.

12 - (x - 1)

Exercise 7:

As you think about each question, fill in the thought bubbles below. Note "*I'm in the mood for pizza,*" is not an acceptable answer.

Suppose you are trying to calculate 18×11 in your head. You decide to use the distributive property and equate 18×11 with $18(10 + 1)$ . . .

= ________

Suppose you are given $15 - (x - 2)$. Which way would you prefer to look at the expression? Choose one way and solve . . .

= ________ = ________

other fun things to do with numbers

absolute value

If you look at a point on the number line, measure its distance from zero, and consider that value as positive, you have just taken the absolute value. Let's take the absolute value of 7.

$|7| = 7$

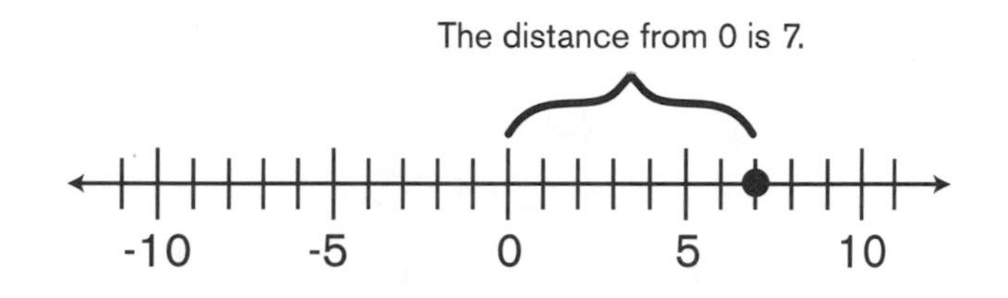

Next, let's calculate $|-7|$, which also equals 7.

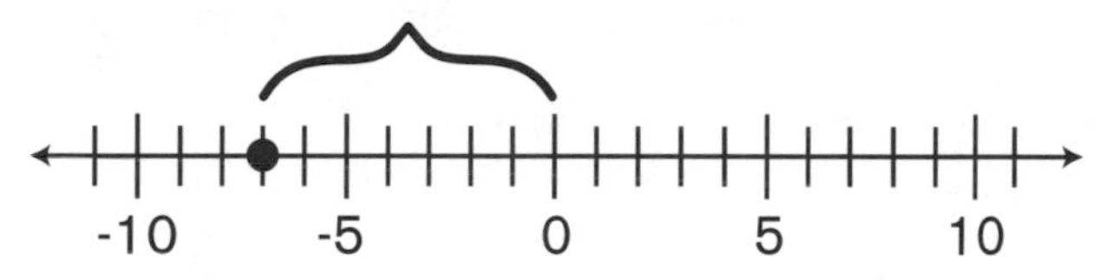

Exercise 8:

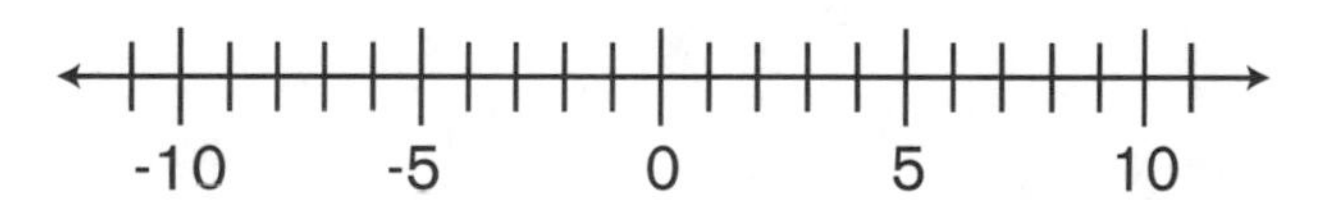

What is $|-11|$?

factorials

When you take the *factorial* of a number, you just multiply that number by every positive whole number less than it. For example, 5 factorial, written **5!** $= 5 \times 4 \times 3 \times 2 \times 1$.

Exercise 9:

8! = ___ × ___ × ___ × ___ × ___ × ___ × ___ × ___

6! = ___ × ___ × ___ × ___ × ___ × ___

3! = ___ × ___ × ___

answers to chapter exercises

Exercise 1:

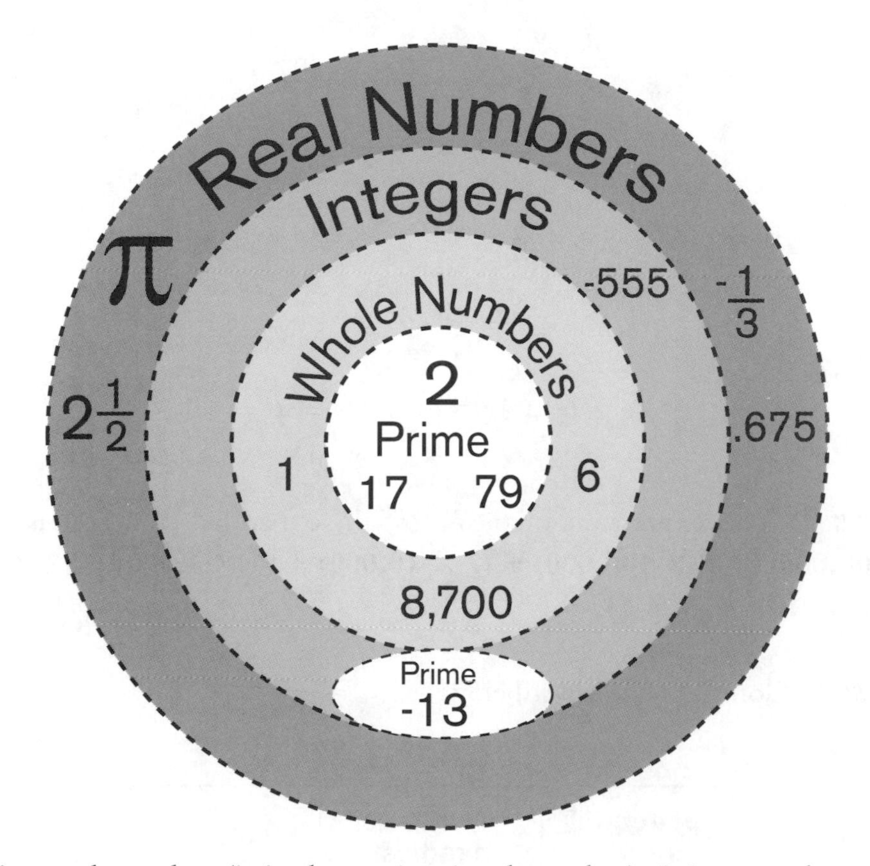

π is a real number (it is also an irrational number). It is a number with a never-ending decimal extension.

$2\frac{1}{2}$, $-\frac{1}{3}$, and .675 are also a real numbers. They are not integers because they involve fractions and decimals.

−13 and −555 are integers (in addition to being real numbers). Notice that −13 is also prime. It cannot go into the center circle labeled "prime" because those prime numbers are also whole numbers. −13 is not a whole number because it is negative.

1, 6, and 8,700 are real numbers, integers, and whole numbers.

The center circle represents numbers that are real, integers, whole, and prime. 2, 17, and 79 can be classified as such.

Exercise 2: Simplify the following.

$$3^2 \bullet 3^4 \bullet 3 = 3^2 \bullet 3^4 \bullet 3^1 = 3^{2+4+1} = 3^7 = 2{,}187$$

$$\frac{7^{10}}{7^5} = 7^{10-5} = 7^5 = 16{,}807$$

$$(2^3b^7)^2 = 2^{3\bullet2}b^{7\bullet2} = 2^6b^{14} = 64b^{14}$$

$$5^2 \div 5^{-2} = 5^{2-(-2)} = 5^{2+2} = 5^4 = 625$$

$$4^2 \bullet 2^5 = (2^2)^2 \bullet 2^5 = 2^4 \bullet 2^5 = 2^{4+5} = 2^9 = 512$$

$$4^{-2} \div 4^{-5} = 4^{-2-(-5)} = 4^{-2+5} = 4^3 = 64$$

Exercise 3: The expression of 6,871,235 in expanded notation is **(6 × 1,000,000) + (8 × 100,000) + (7 × 10,000) + (1 × 1,000) + (2 × 100) + (3 × 10) + (5 × 1).**

Exercise 4: Complete the chart below.

exponential form	rewrite as a radical	solve
$125^{\frac{1}{3}}$	$\sqrt[3]{125}$	$= 5$
$121^{\frac{1}{2}}$	$\sqrt[2]{121}$ $\sqrt{121}$	$= 11$
$64^{\frac{1}{6}}$	$\sqrt[6]{64}$	$= 2$
$-8^{\frac{1}{3}}$	$\sqrt[3]{-8}$	$= -2$

Exercise 5:

$2^3 - 7 \times (5 - 8) \div 3$	
Parentheses	$2^3 - 7 \times (5 - 8) \div 3$ $2^3 - 7 \times (-3) \div 3$ $2^3 - 7 \times -3 \div 3$
Exponents	$2^3 - 7 \times -3 \div 3$ $8 - 7 \times -3 \div 3$
Multiplication/Division	You do the multiplication first because it appears first . . . $8 - 7 \times -3 \div 3$ $8 - (-21) \div 3$ Next, you do the division . . . $8 - (-21) \div 3$ $8 - (-7)$
Addition/Subtraction	$8 - (-7)$ $8 + {}^{+}7$ 15

Exercise 6:

"10 to what power is 10,000?"

$10^4 = 10{,}000$, so the answer is 4.

"10 to what power is 1000?"

$10^3 = 1{,}000$, so the answer is 3

Exercise 7:

18 times 10
plus 18
times 1.

$$18(10 + 1)$$

$$= (18 \times 10) + (18 \times 1)$$

$$= 180 + 18$$

$$= 198$$

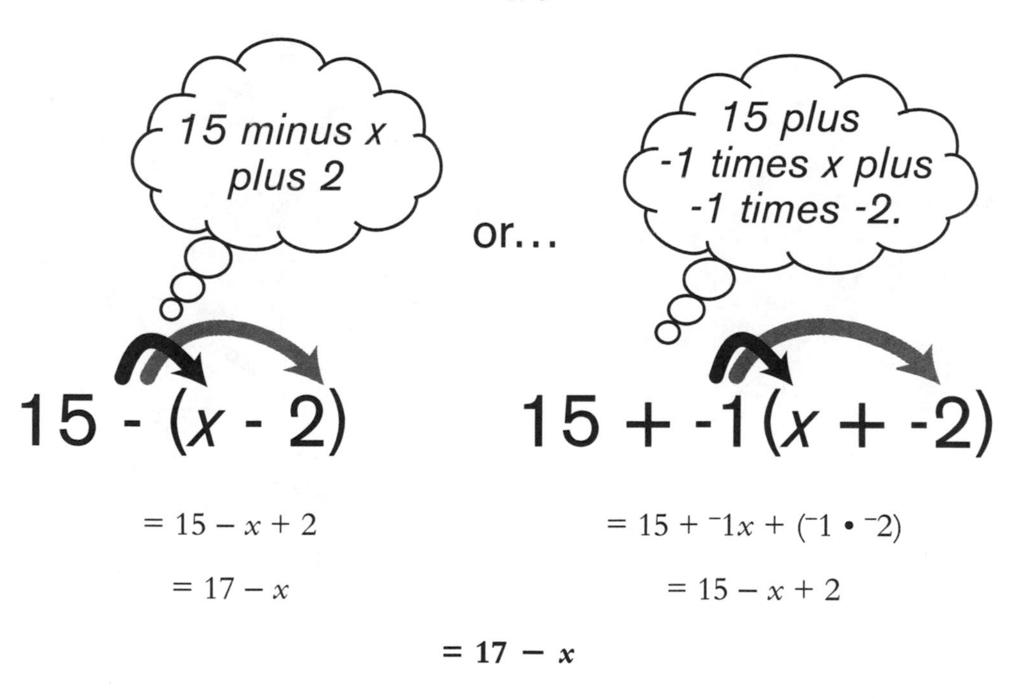

Exercise 8: On the number line below, mark the point −11.

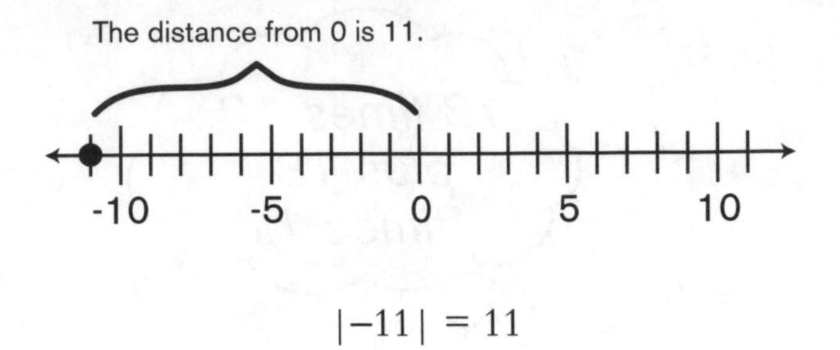

$$|-11| = 11$$

Exercise 9:

$$8! = 8 \times 7 \times 6 \times 5 \times 4 \times 3 \times 2 \times 1$$

$$6! = 6 \times 5 \times 4 \times 3 \times 2 \times 1$$

$$3! = 3 \times 2 \times 1$$

Fractions and Decimals

introduction to fractions

Fractions are used to represent parts of a whole. You can think of the fraction bar as meaning "*out of*" . . .

You can also think of the fraction bar as meaning "*divided by*" . . .

examples of fractions:

 One out of two pieces is filled in . . .

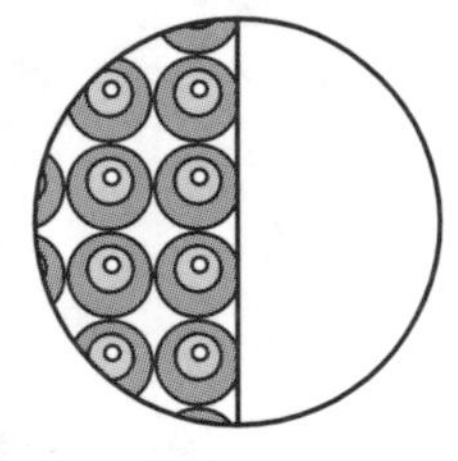

 One out of three pieces is shaded . . .

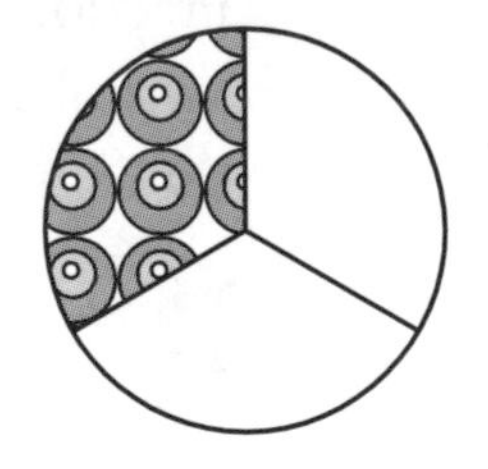

$\frac{1}{4}$ One out of four pieces is shaded . . .

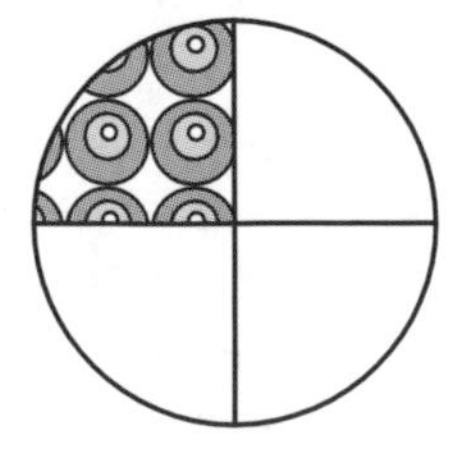

$\frac{1}{12}$ One out of twelve pieces is shaded . . .

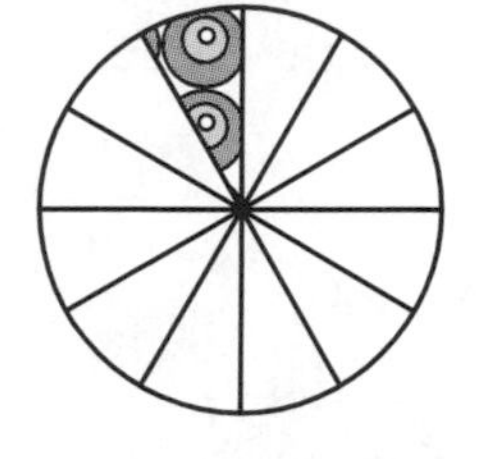

Although most people call the top part of the fraction the "*top*" and the bottom part of a fraction the "*bottom*," the technical names are **numerator** and **denominator**.

You are never allowed to have a zero in the denominator. Anything divided by zero is undefined.

A **proper fraction** has a numerator that is smaller than its denominator. Examples are $\frac{1}{2}$, $\frac{13}{25}$, and $\frac{99}{100}$. **Improper fractions** have numerators that are bigger than their denominators. Examples include $\frac{8}{5}$, $\frac{99}{13}$, and $\frac{5}{2}$.

A fraction that represents a particular part of the whole is sometimes referred to as a **fractional part**. For example, let's say that a family has four cats and two dogs. What fractional part of their pets are cats?

Since four out of the total six animals are cats, the fractional part of their pets that are cats is equal to $\frac{4}{6}$. You can **reduce** this fraction to $\frac{2}{3}$: $\frac{4 \div 2}{6 \div 2} = \frac{2}{3}$

What fractional part of the figure below is shaded?

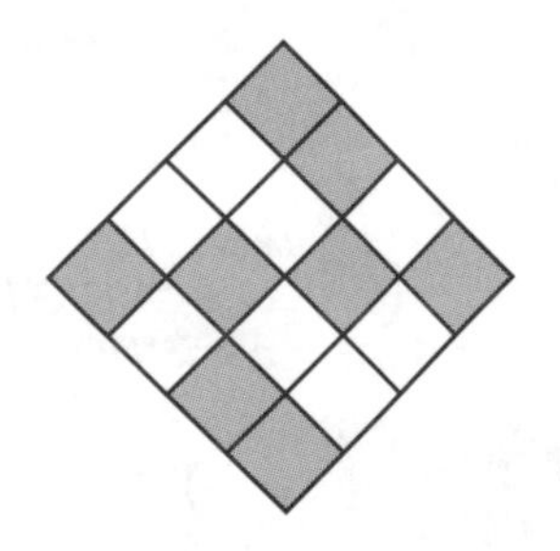

Let's rearrange the shaded areas . . .

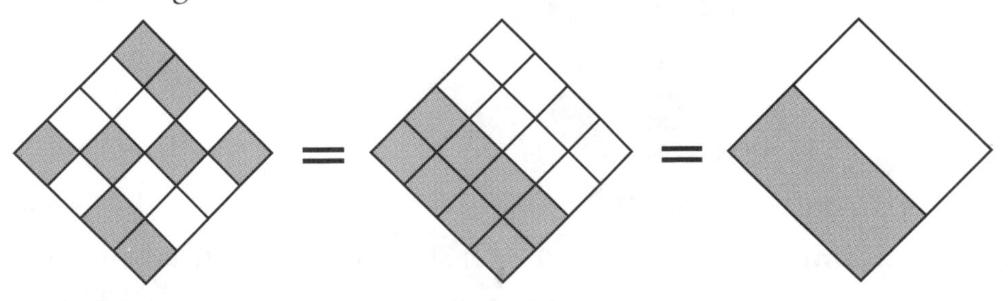

Thus, $\frac{1}{2}$ is shaded.

Exercise 1: Use the data below to answer the following questions:

- What fractional part of the days listed were sunny?
- What fraction of the days were 40°?
- What fraction of the days were 55° or higher?
- What fractional part of the days listed were greater than 60°?

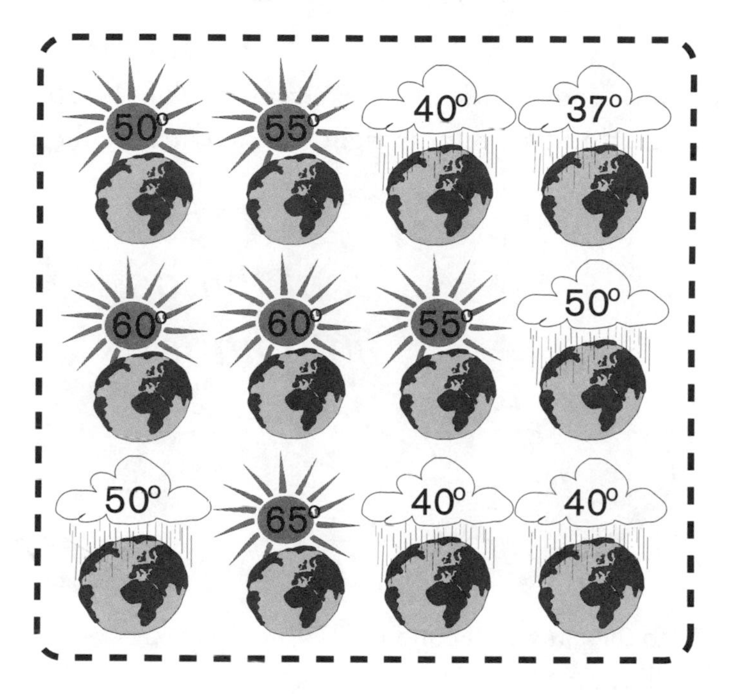

adding and subtracting fractions

In order to add or subtract fractions, you must have the same denominator.

$$\frac{3}{11} + \frac{4}{11} = \frac{\text{numerator} + \text{numerator}}{\text{denominator}} = \frac{7}{11}.$$

When you don't have the same denominator you can convert the denominators so that you can add or subtract. You should find the **least common denominator (LCD)**. Think of the least common denominator as a cookie cutter. Let's say you are adding $\frac{1}{3}$ and $\frac{1}{4}$ of a cookie.

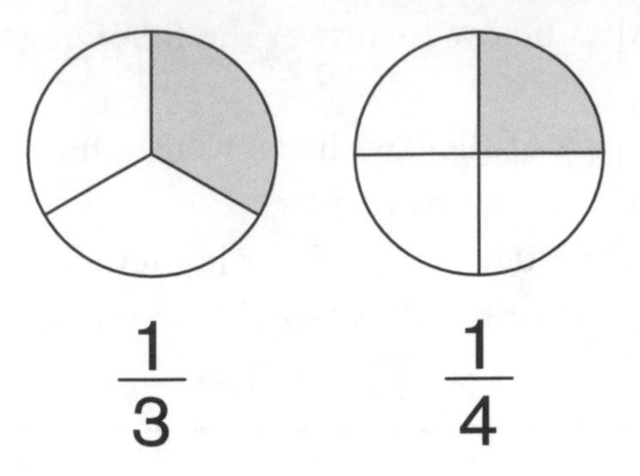

You can use your cookie cutter to make both cookies have the same sized slices. It is important to pick out the right cookie cutter. Otherwise, you get a mess . . .

If you just grab the *fifths* cutter and cut both our cookies, you get something that looks like this . . .

Let's try that again. You have $\frac{1}{3}$ and $\frac{1}{4}$. The **least common multiple (LCM)** of 3 and 4 is 12. Let's try the *twelfths* cutter . . .

You will cut both cookies into 12 slices . . .

Now you add $\frac{4}{12} + \frac{3}{12}$ to get $\frac{7}{12}$. Note that once you pick out the new denominator that you need, you can multiply your fraction by a version of "1" and that gives you a new denominator.

$$\frac{1}{3} \times \text{“1”} = \frac{1}{3} \times \frac{4}{4} = \frac{4}{12}$$

$$\frac{1}{4} \times \text{“1”} = \frac{1}{4} \times \frac{3}{3} = \frac{3}{12}$$

converting improper fractions into mixed numbers

Let's add $\frac{3}{4}$ and $\frac{5}{8}$ . . .

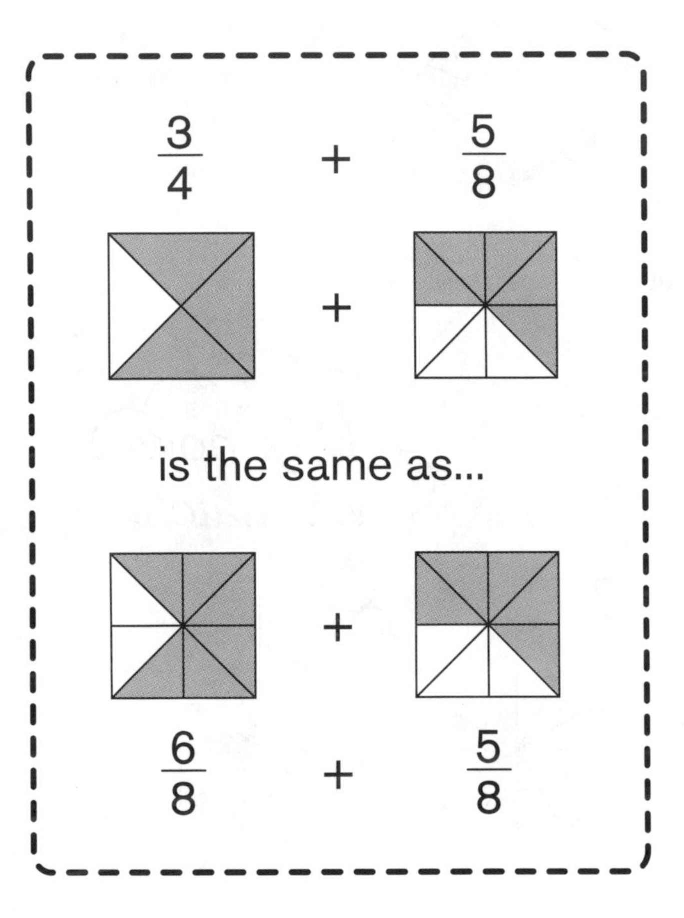

$\frac{6}{8} + \frac{5}{8} = \frac{11}{8}$, which is an *improper fraction*. To convert $\frac{11}{8}$ to a **mixed number**, you just divide 11 by 8 to get 1 with a remainder of 3. Since you're dealing with eighths, you stick the remainder over 8, yielding $1\frac{3}{8}$. If you look at the two fractions below, you can see how easily $\frac{6}{8} + \frac{5}{8}$ can be rearranged into a *whole* with three slices left over.

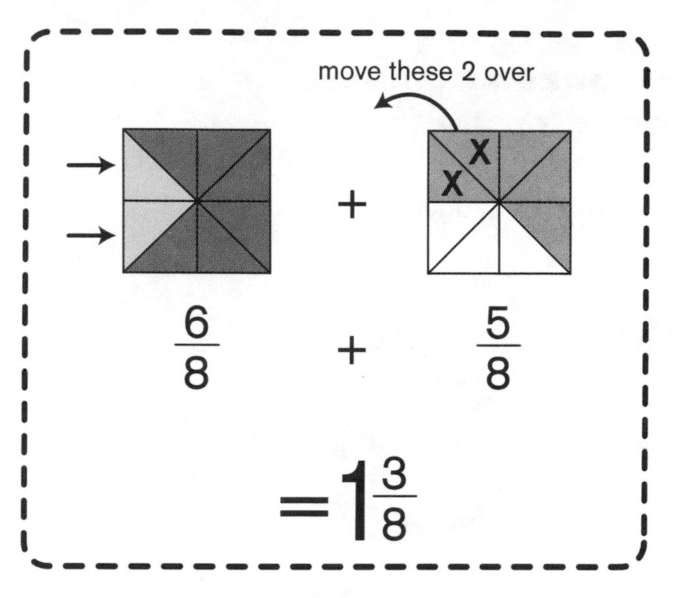

Exercise 2:

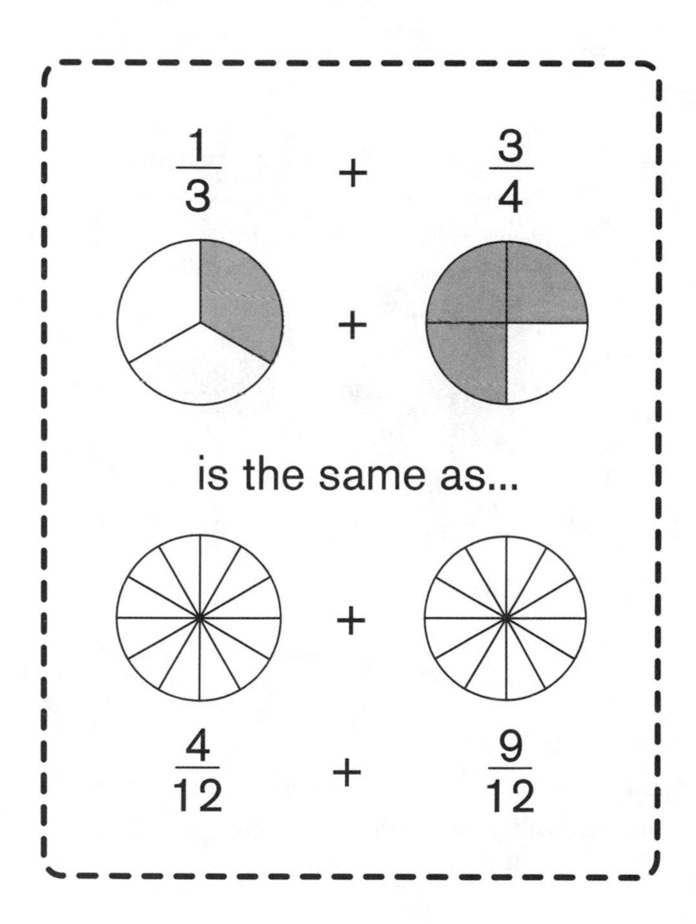

converting mixed numbers into improper fractions

Let's consider the mixed number $1\frac{1}{4}$.

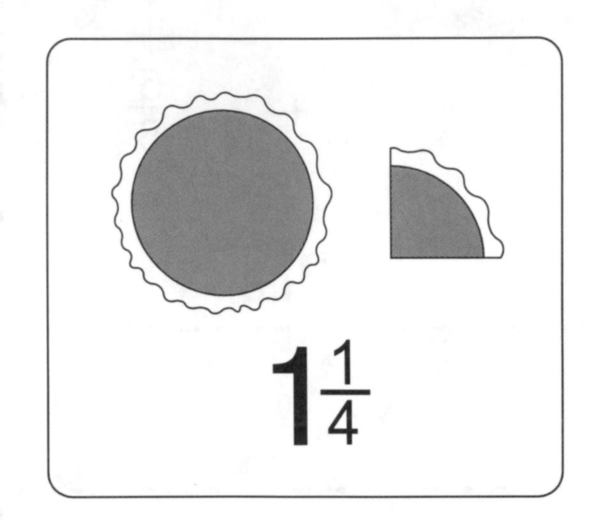

You could easily cut the whole pie into four quarters . . .

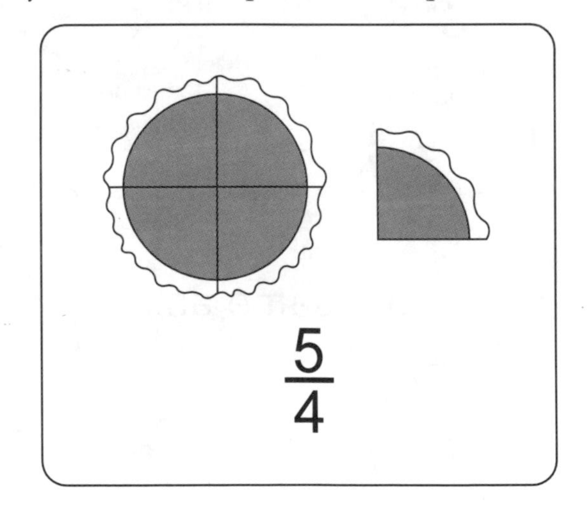

Thus, you know that $1\frac{1}{4} = \frac{5}{4}$.

To convert mixed numbers into improper fractions on the fly, you just multiply the whole number by the denominator, add this to the numerator, and stick this value over the same denominator . . .

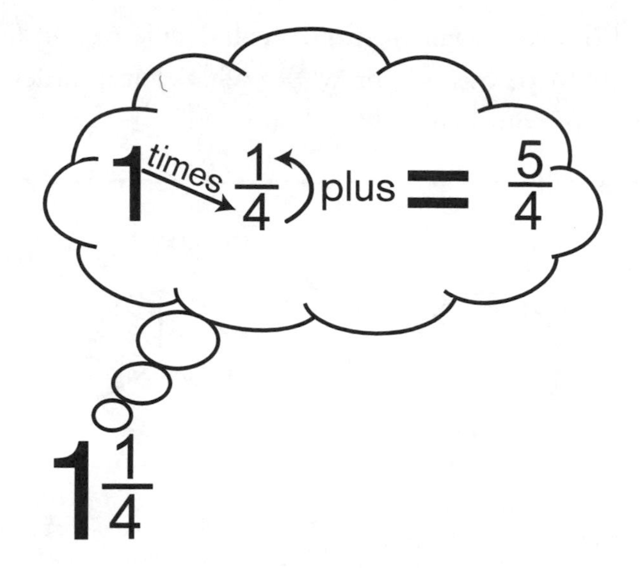

multiplying fractions

You don't need to worry about lowest common denominators (LCDs) or lowest common multiples (LCMs) when you multiply fractions. You just multiply the numerators and the denominators:

$$\frac{1}{5} \times \frac{3}{5} = \frac{\text{numerator} \times \text{numerator}}{\text{denominator} \times \text{denominator}} = \frac{1 \times 3}{5 \times 5} = \frac{3}{25}$$

Sometimes, you can reduce before you multiply:

$$\frac{2}{3} \times \frac{3}{5} =$$

$$\frac{2}{\underset{1}{\cancel{3}}} \times \frac{\overset{1}{\cancel{3}}}{5} = \frac{2}{5}$$

If you are asked to find the fraction **of** a number, just **multiply** that number by the fraction:

of means ×

$\frac{1}{3}$ ***of*** 9 means $\frac{1}{3} \times 9 = 3$.

Suppose that from your available funds, you decide to transfer $\frac{1}{3}$ of it into a checking account. Next, you withdraw $\frac{1}{4}$ of what's left in order to buy a computer. How much of your original funds are left?

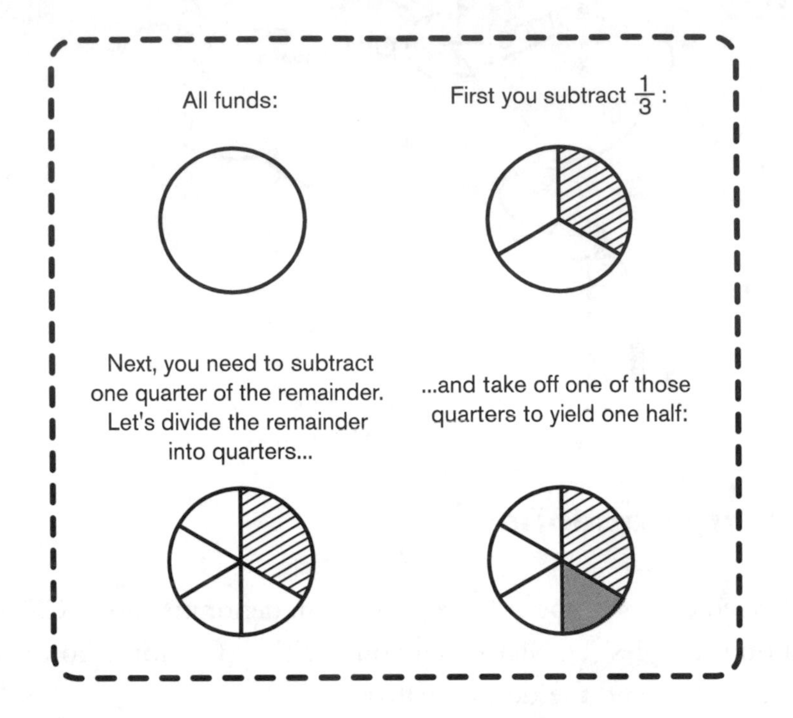

Half your funds are left. So if you started with $6,000, you'd have $3,000 left. If you started with $10,000, you'd have $5,000 left.

You can look at the same question in another way. Notice that the fractions you will be working with are $\frac{1}{3}$ and $\frac{1}{4}$. Dividing the *whole* into 3×4, or 12 wedges will guarantee that you can take all the fractional parts away neatly.

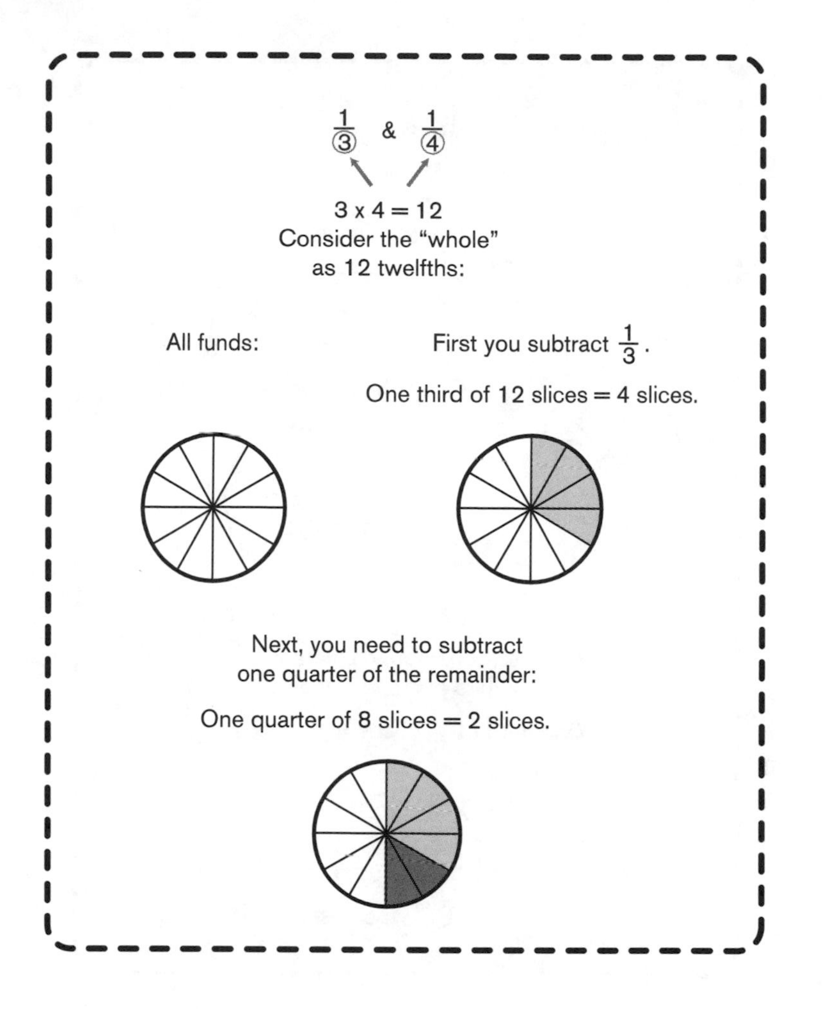

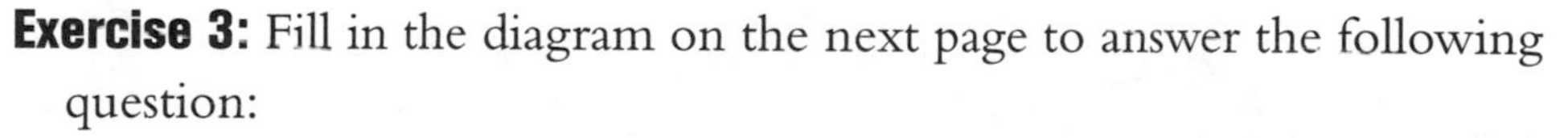

Exercise 3: Fill in the diagram on the next page to answer the following question:

Fifteen new children just enrolled at a day camp. First, $\frac{1}{5}$ of the children are placed into the preschool group. Next, $\frac{1}{3}$ of the remaining campers are placed into the girls group. What fractional part of the new campers has yet to be placed?

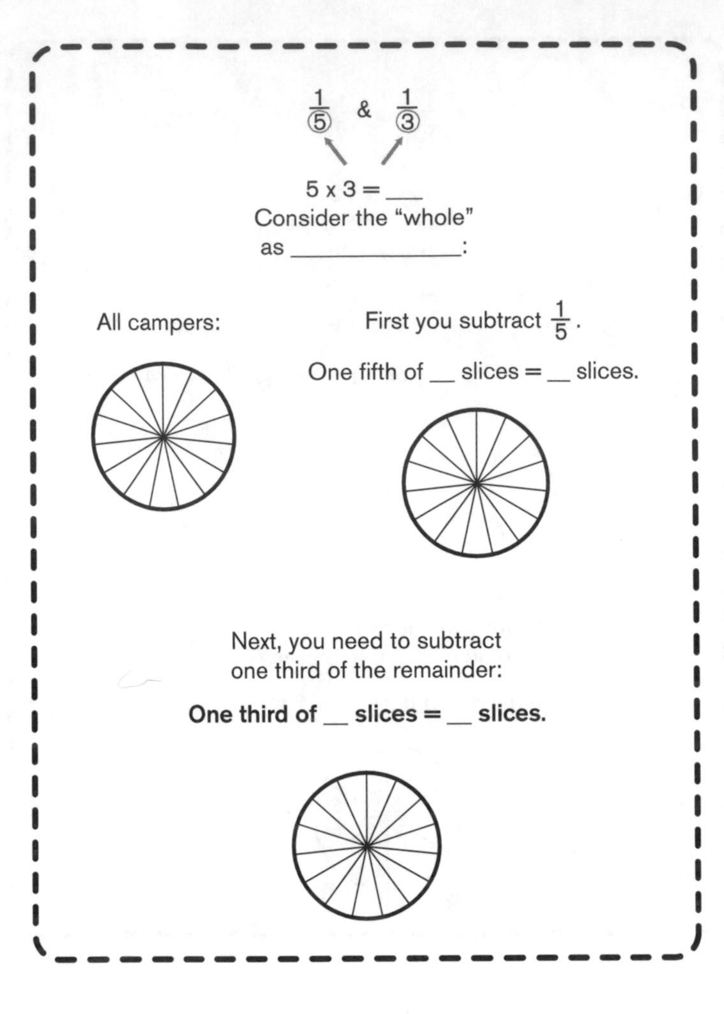

dividing fractions

Pretend you had $\frac{1}{2}$ a pie and you wanted to share it with somebody. It is clear that you would each get $\frac{1}{4}$ of the pie (that is, if you were being *fair*).

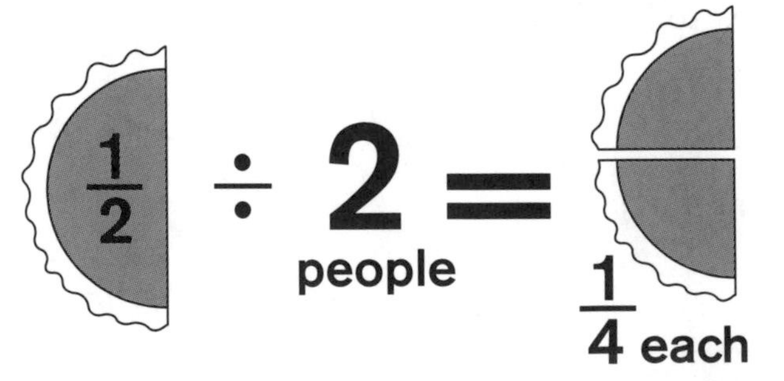

What if you had two donuts and lots of mouths to feed? You might decide to cut the donuts into quarters so that you could divvy them out . . .

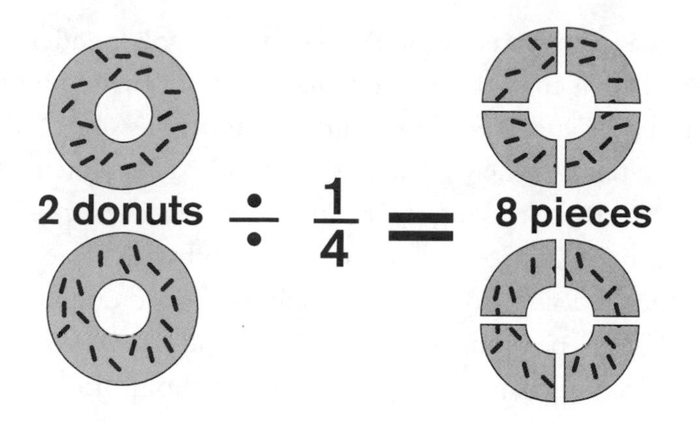

You would then be able to feed eight mouths.

You may remember that in order to divide one fraction by another, you need to flip the second fraction and then *multiply* the fractions:

$$\frac{3}{5} \div \frac{2}{3} =$$

First, take the **reciprocal** of $\frac{2}{3}$ . . .

Then, multiply: $\frac{3}{5} \times \frac{3}{2} = \frac{9}{10}$

Exercise 4: Look back at the pie example and the donut example and rewrite the division problems as multiplication problems:

- $\frac{1}{2} \div 2 =$
- $2 \div \frac{1}{4} =$

fraction word problems

Your friend is an aspiring screenwriter and comes to you for advice. She is writing a scene in which a character named Dominick is obsessed with listening to his Yanni album over and over again. In desperation, his sister hides the record under a couch cushion when he leaves for work. Later, their mutual friend Alice sits on the couch, breaking the album into 8 equal pieces. To make matters worse, their neighbor Joey's hamster makes off with a piece and eats $\frac{2}{3}$ of it. When Dominick is apprised of the ill fate his record endured, he demands that Joey pay $12 for a new one. However, Joey replies that he should only have to pay for the piece his hamster ate. How much should Joey claim he owes?

Joey's hamster ate $\frac{2}{3}$ of $\frac{1}{8}$. $\frac{2}{3} \times \frac{1}{8} = \frac{2}{24}$, or $\frac{1}{12}$. This means that he is financially obligated to pay $\frac{1}{12}$ of the \$12. $\frac{1}{12}$ of \$12 $= \frac{1}{12} \times \$12 = \$1$.

Exercise 5: Athena is wallpapering one wall of her room. The following diagram shows how much she has completed so far. She already used up

$15 worth of wallpaper. Estimate how much <u>more</u> it will cost her to finish the wall.

introduction to decimals

Decimals are really just fractions in disguise. Or maybe it's the fractions that are really decimals in disguise. You may never know who's disguised as what in this crazy world of mathematical espionage. One thing is for sure; you can equate fractions and decimals. For example, the fraction $\frac{1}{10}$ can be written as the decimal .1

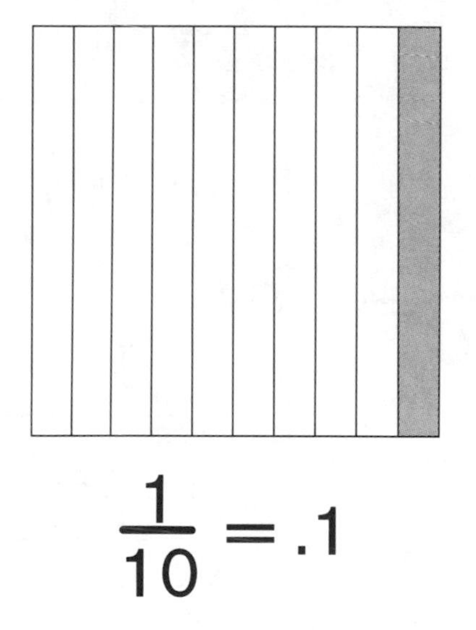

$$\frac{1}{10} = .1$$

The fraction $\frac{1}{100}$ is equal $1 \div 100 = .01$

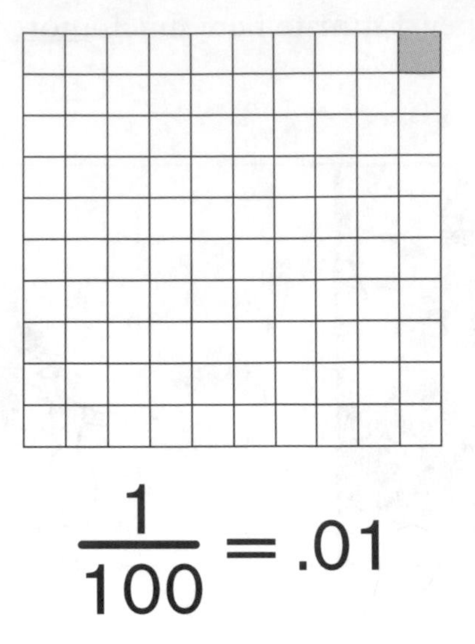

$$\frac{1}{100} = .01$$

The fraction $\frac{1}{1000}$ is equal $1 \div 1000 = .001$

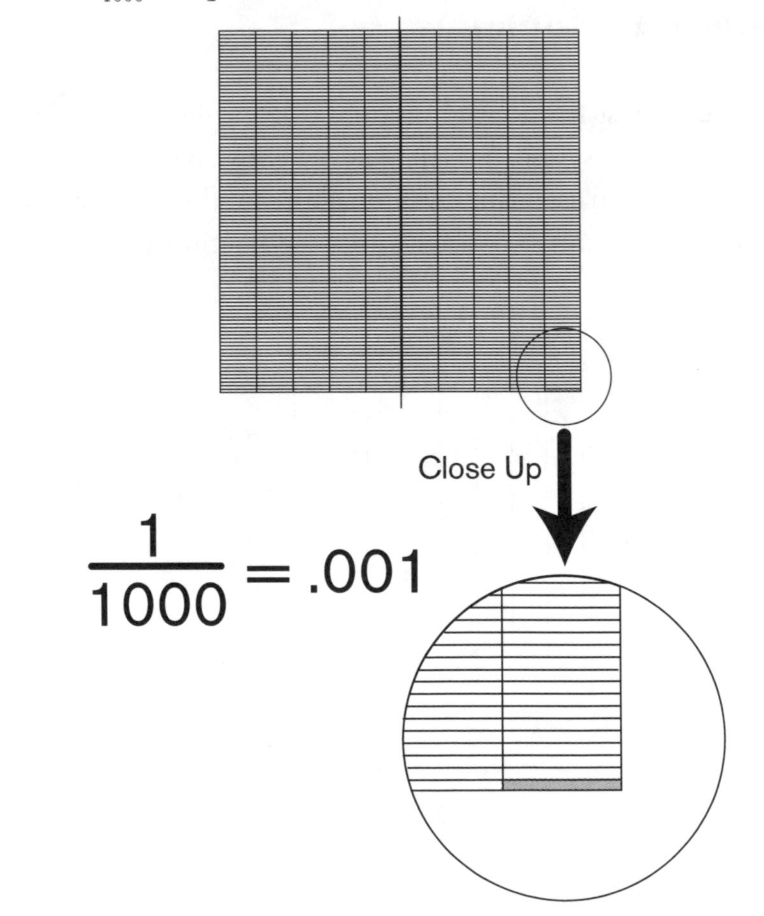

$$\frac{1}{1000} = .001$$

Some decimals can only approximate fractions because they are *irrational numbers*. For example, π, $\sqrt{2}$, and $\sqrt{11}$ are equivalent to decimals that never terminate or repeat. Sometimes we approximate π as $\frac{22}{7}$.

$$\pi = 3.141592\ldots$$

$$\sqrt{2} = 1.414213\ldots$$

$$\sqrt{11} = 3.316624\ldots$$

place value

You already looked at the names of the places to the left of the decimal point in Chapter 1. Now, you will look at the names of the places to the right of the decimal point.

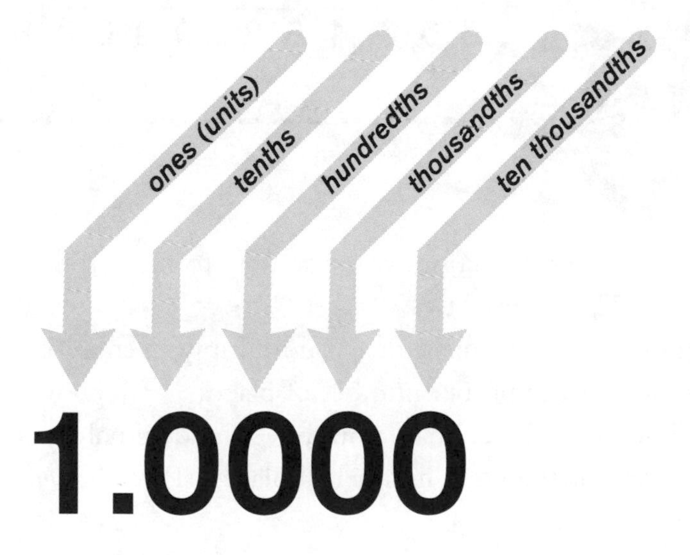

operations with decimals

The following diagram reviews basic operations involving decimals. These are the types of calculations you need to make when dealing with money, such as when balancing your checkbook.

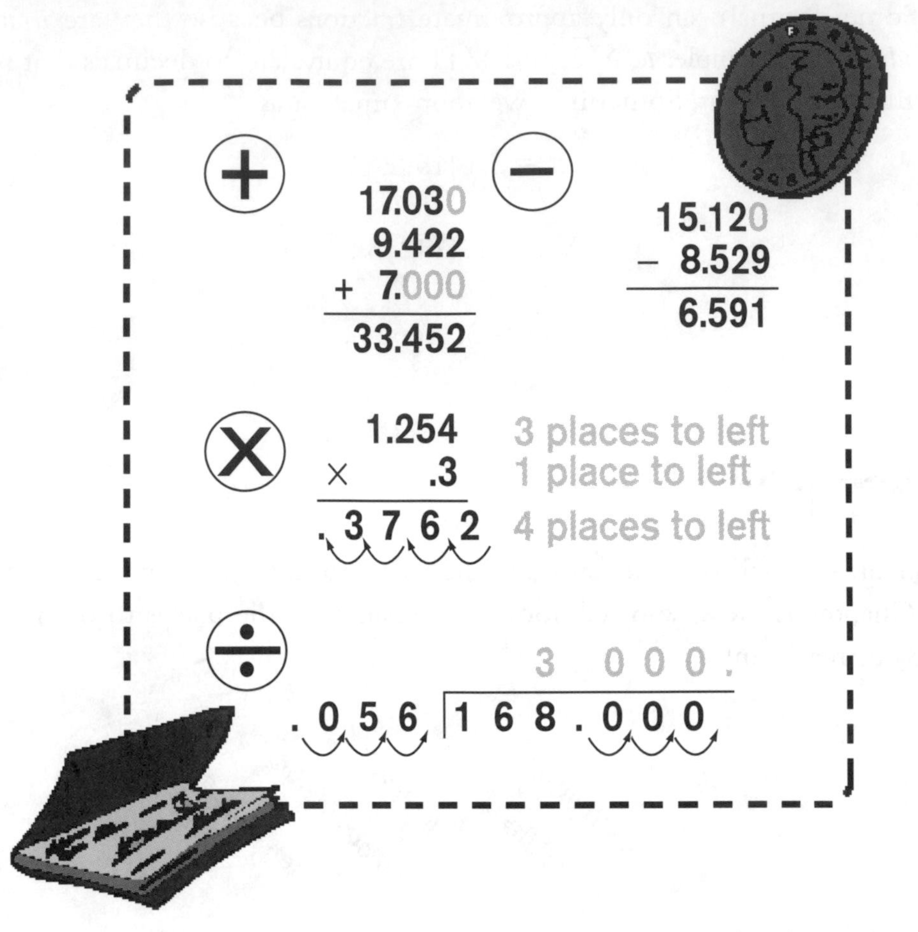

- For addition and subtraction, you just need to line up the decimal points and add or subtract.
- When multiplying decimals, first you multiply in the usual fashion, and then count over the proper number of places.
- When dividing with decimals, you move the decimal point of the dividend and divisor the same number of places. (Recall: *dividend* ÷ *divisor* = *quotient*.)

converting fractions to decimals

To convert a fraction to a decimal, just divide the top by the bottom. For example, $\frac{1}{2}$ would equal $1 \div 2$, or .5.

Memorizing some of the decimal values of the more common fractions can come in handy:

Fraction	Decimal
$\frac{1}{2}$	.5
$\frac{1}{3}$	.3
$\frac{2}{3}$	.6
$\frac{1}{4}$	.25
$\frac{3}{4}$	.75
$\frac{1}{5}$	.2

converting decimals to fractions

If you can say the name of the decimal, you can easily convert it to a fraction. For example, if you see .123, you can say "*One hundred and twenty-three thousandths*," which is the same as $\frac{123}{1000}$.

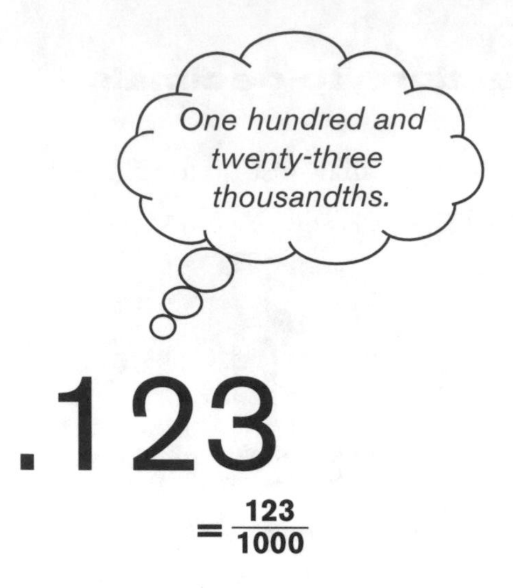

scientific notation

In the old days scientists found themselves dealing with big, big numbers like 13,000,000,000 and really little numbers like .000000013. They were very sad because they did not like having to write so many zeros. It hurt their hands, wasted chalk, and took time away from playing practical jokes on each other.

Then one day, a scientist came up with the idea to use "shorthand" for all those zeros. The "shorthand" incorporates the powers of ten . . .

10^0	= 1	
10^1	= 10	
10^2	= 100	
10^3	= 1,000	
10^4	= 10,000	
10^5	= 100,000	
10^6	= 1,000,000	*one million*
10^7	= 10,000,000	
10^8	= 100,000,000	
10^9	= 1,000,000,000	*one billion*
10^{10}	= 10,000,000,000	
10^{11}	= 100,000,000,000	
10^{12}	= 1,000,000,000,000	*one trillion*
10^{13}	= 10,000,000,000,000	
10^{14}	= 100,000,000,000,000	

This meant that instead of writing 13,000,000,000, one could write 1.3×10^{10}. Instead of writing .000000013, one could write 1.3×10^{-8}. Then all the scientists were very happy and called this shorthand **scientific notation**.

Fortunately, you do not have to memorize the chart above. The trick to expressing, say, 250,000,000,000,000 as 2.5 times a power of 10 is to start at the current decimal point and then count until you reach the place where you want to insert the new decimal point.

250,000,000,000,000.

You counted 14 places to the left, so 250,000,000,000,000 = 2.5×10^{14}.

Let's divide $7.6 \div 10^4$ by 200:

$$\frac{7.6 \times 10^4}{200} = .038 \times 10^4$$

The standard form dictates that you express $.038 \times 10^4$ as 3.8 times a power of 10. But what power of ten? Note that the 10^4 is really a bunch of powers of ten neatly stored in exponential form. You can **steal** some powers of 10 in order to move our decimal over to the **right**.

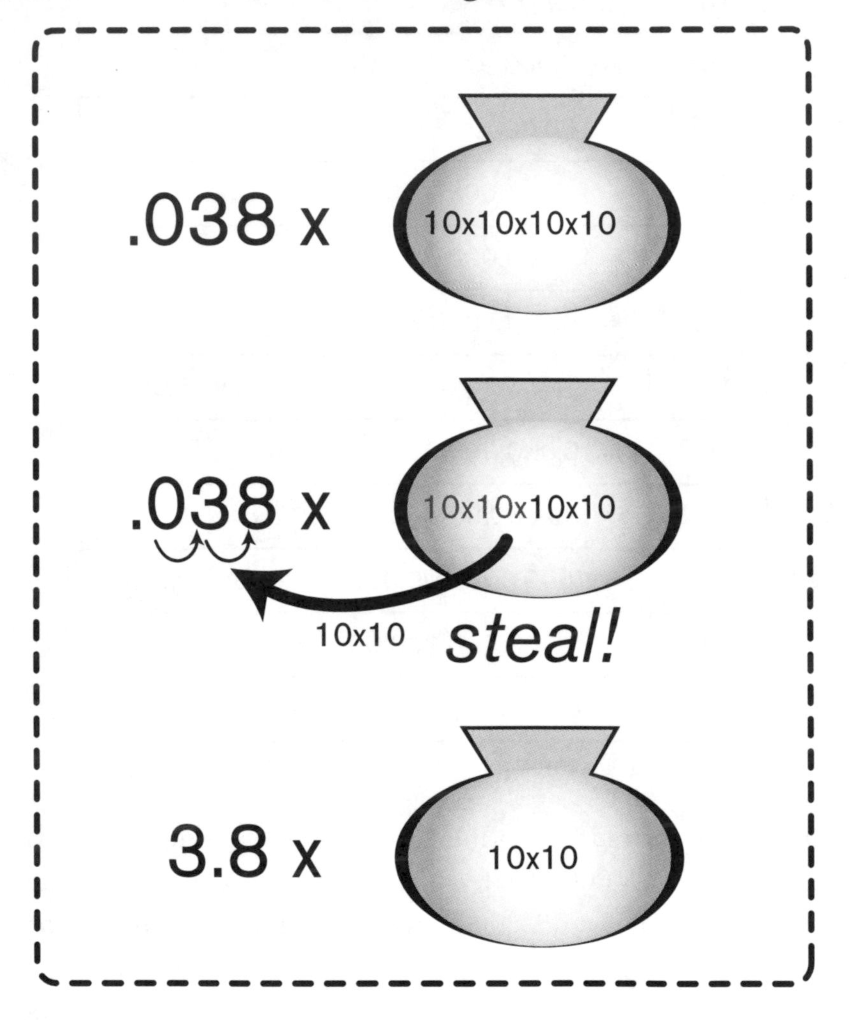

Notice that when you *stole* the 10×10, you **subtracted** from the exponent in 10^4, effectively turning $.038 \times 10^4$ into 3.8×10^2. Similarly, $5{,}000 \times 10^4$ becomes 5×10^7.

Sometimes you may have to move the decimal point to the left. Let's convert $5{,}000 \times 10^4$ into standard scientific notation format.

First, notice that $5{,}000 = 5 \times 1{,}000$.

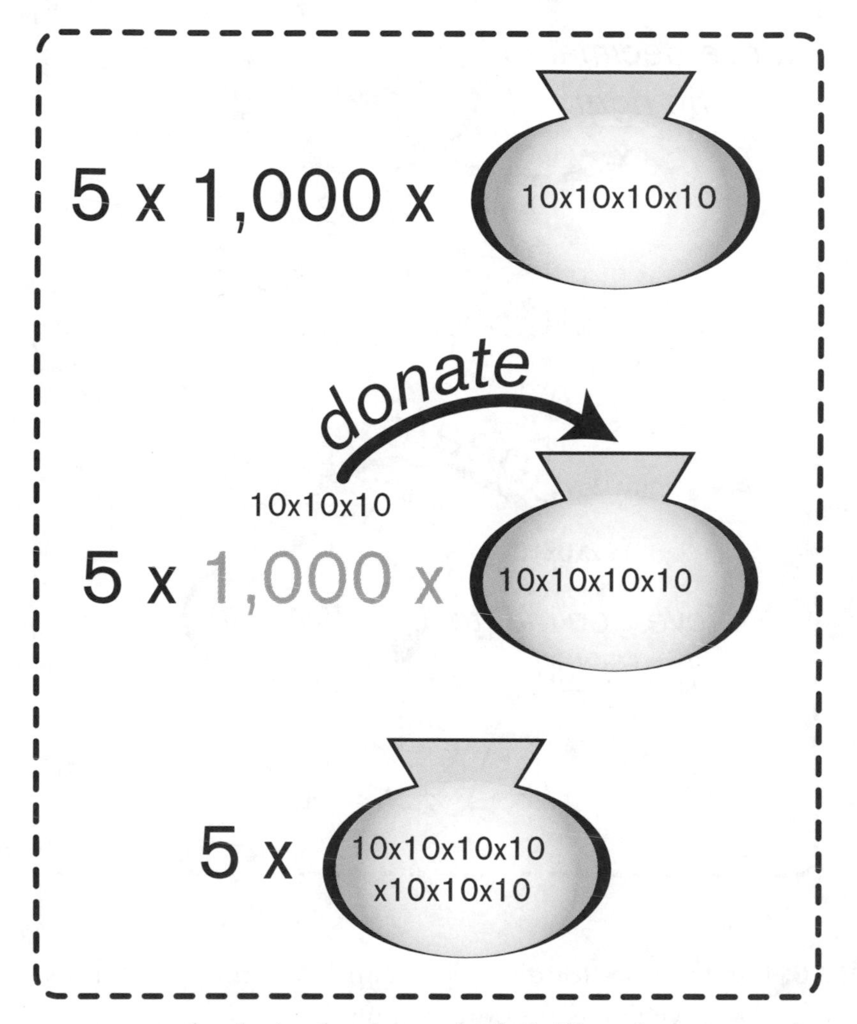

When you move the decimal point to the **left** (like when you go from 5,000 to 5.0) you **donate** some powers of 10, and you **add** to the exponent.

To summarize:

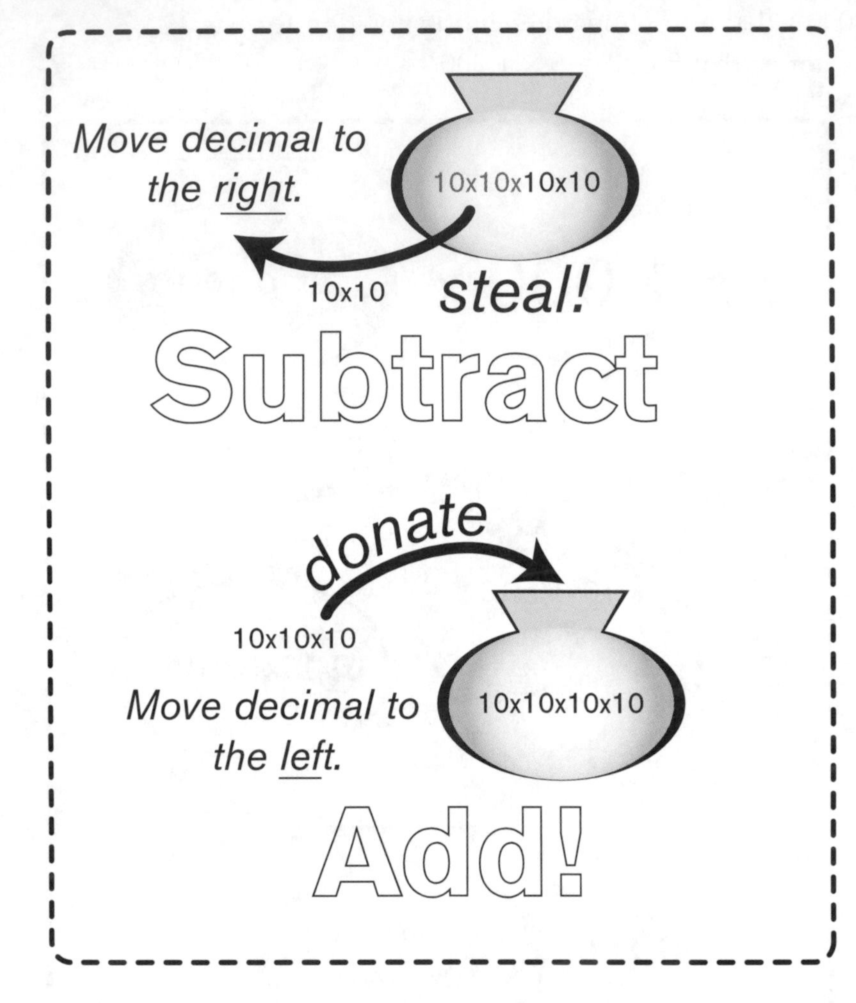

Exercise 6: Use the rules for dividing decimals and the rules for dividing exponents to solve the expressions below:

$$\frac{6.3 \times 10^4}{2.1 \times 10^8} =$$

- Express the answer in scientific notation. ________
- Express the answer as a decimal. ________

Hint: *Divide the 6.3 by the 2.1 and the 10^4 by the 10^8 and then convert to standard form if necessary.*

Use the following diagram to convert $6{,}500 \times 10^6$ into standard scientific notation form.

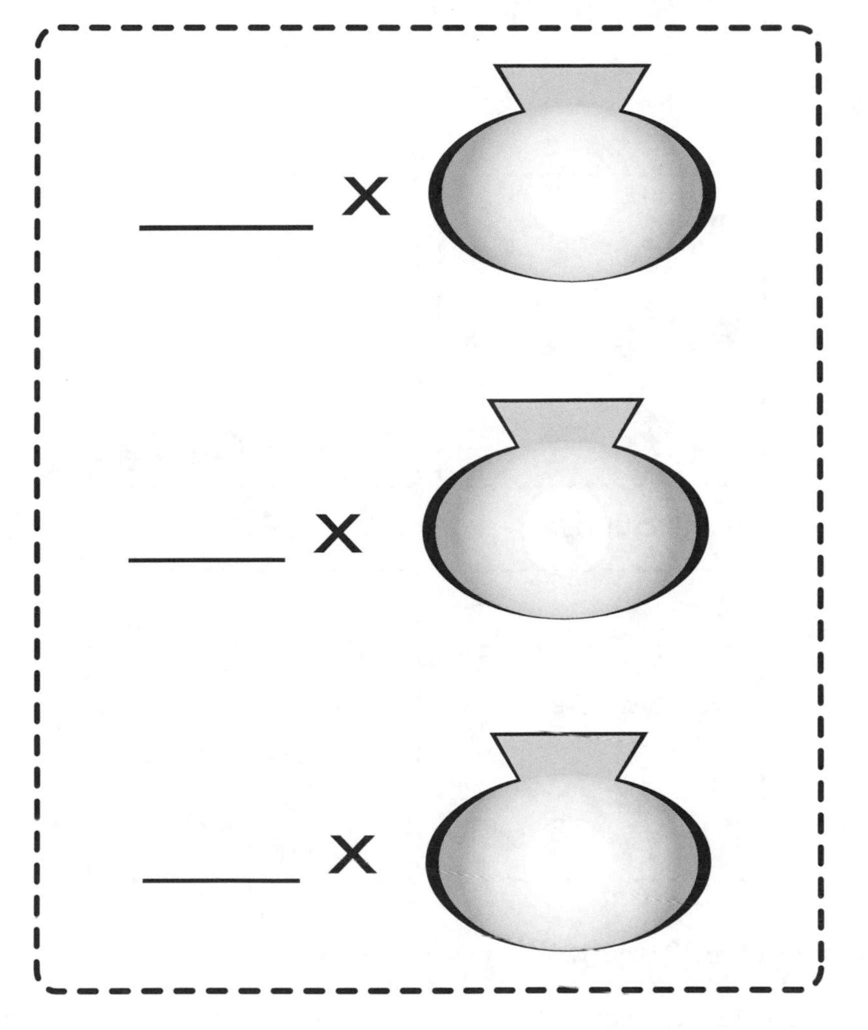

solutions to chapter exercises

Exercise 1:

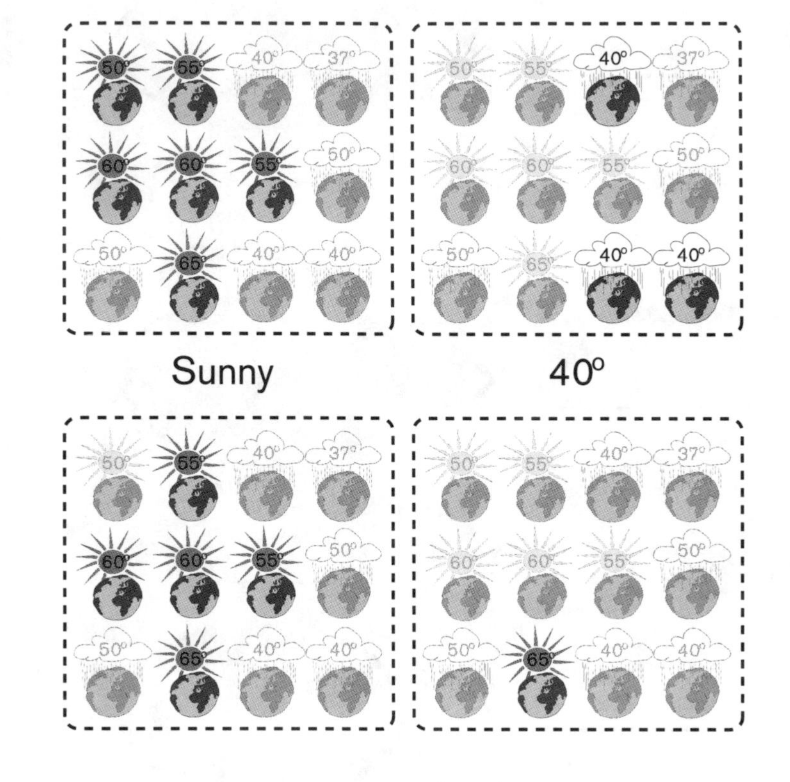

- 6 out of the 12 days were sunny. $\frac{6}{12} = \frac{1}{2}$
- 3 out of 12 days were 40°. $\frac{3}{12} = \frac{1}{4}$
- 5 out of 12 were 55° or higher. $\frac{5}{12}$
- Only one day was the temperature greater than 60°, so the fractional part is $\frac{1}{12}$.

Exercise 2:

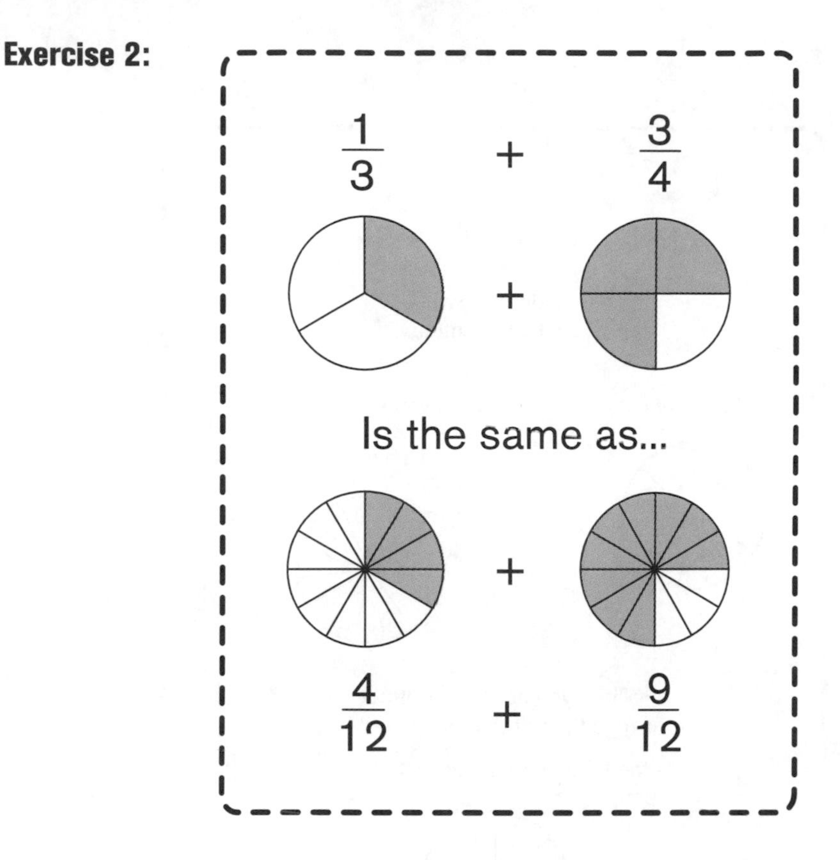

$\frac{4}{12} + \frac{9}{12} = \frac{13}{12}$. $13 \div 12 = 1$ with a remainder of 1. Since you're dealing with twelfths, you stick the remainder over 12, making the answer $1\frac{1}{12}$. Note that in the diagram, you could also move 3 slices from the $\frac{4}{12}$ and turn the $\frac{9}{12}$ into a *whole*, or 1. You would then have 1 whole and 1 slice, or $1\frac{1}{12}$:

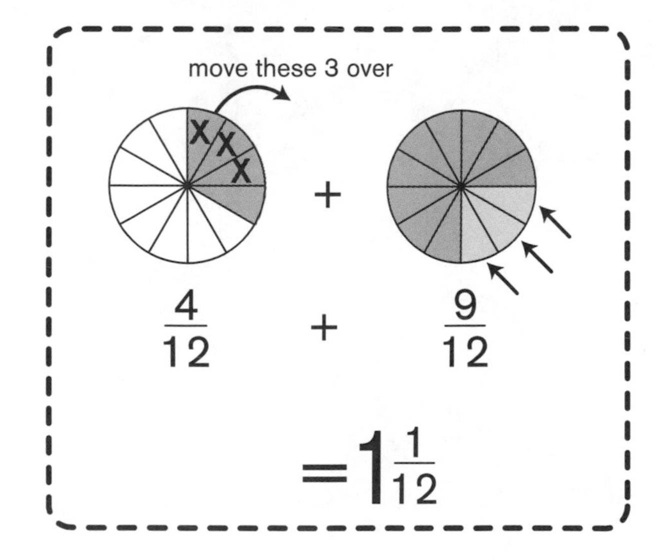

Exercise 3:

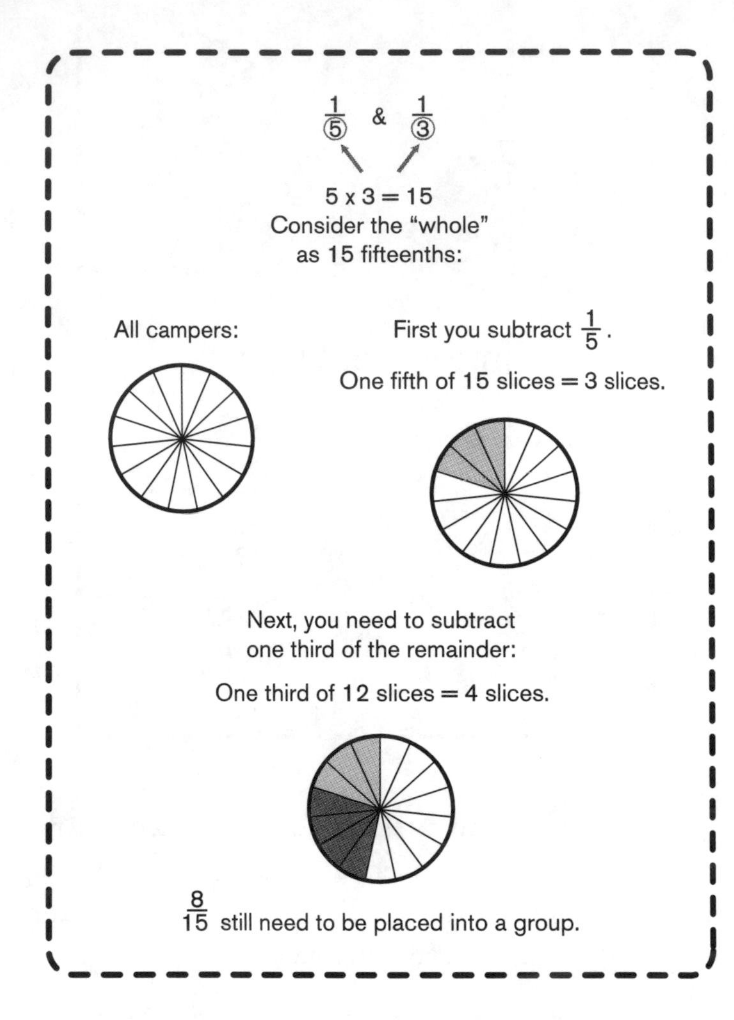

Exercise 4:

- $\frac{1}{2} \div 2 = \frac{1}{2} \div \frac{2}{1} = \frac{1}{2} \times \frac{1}{2} = \frac{1}{4}$
- $2 \div \frac{1}{4} = 2 \times \frac{4}{1} = 8$

Exercise 5: First, estimate how much of the wall has been wallpapered so far.

She has completed $\frac{1}{3}$ and it cost her \$15. To complete each additional third, it will cost \$15. Since there are two-thirds left that need coverage, she will pay an additional $2 \times \$15 = \$30$.

Exercise 6: $\frac{6.3 \times 10^4}{2.1 \times 10^8} = \frac{6.3}{2.1} \times \frac{10^4}{10^8}$

- $6.3 \div 2.1 = 3$, and $10^4 \div 10^8 = 10^{4-8} = 10^{-4}$, so you have 3×10^{-4}. This answer is already in the standard form of scientific notation.
- $3 \times 10^{-4} = 3 \times \frac{1}{10^4} = 3 \times \frac{1}{10,000} =$ *3 ten thousandths* $= .0003$.

To convert 3×10^{-4} to a decimal, move the decimal 4 places to the left.

.0 0 0 3.

(move 4 places to the left)

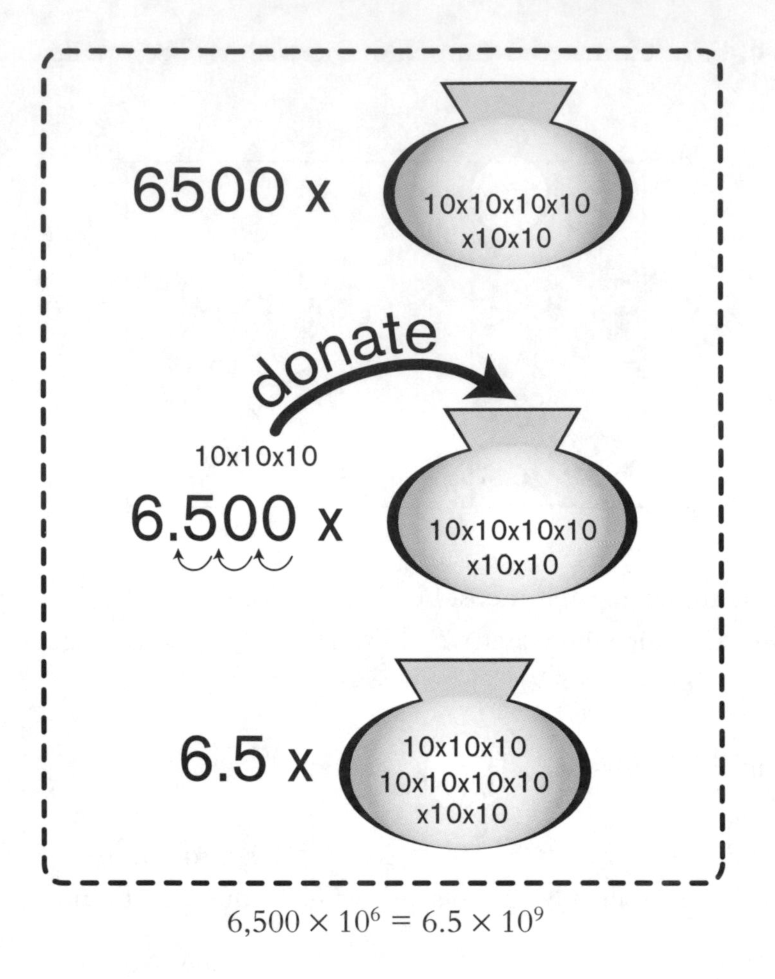

$$6{,}500 \times 10^6 = 6.5 \times 10^9$$

You moved the decimal 3 places to the left and added 3 to the exponent.

chapter
three

Ratios and Proportions

ratios

Ratios are a way of comparing numbers. If your basketball team has 3 boys for every girl, you can express this as a 3 to 1 ratio. You can express ratios in three ways:

- In sentence form:
  There is a three to one ratio of boys to girls.
- By using a colon: 3:1
- By using a bar: $\frac{3}{1}$

Your team:

Just like fractions, ratios can be reduced to smaller terms. Someone may count that your team has 12 boys and 4 girls. That person may comment that there is a 12 to 4 ratio of boys to girls. This is true because mathematically $\frac{12}{4}$ reduces to $\frac{3}{1}$.

Notice that when dealing with a 3:1 ratio, the total is going to be a multiple of 3 + 1, or 4. Here you have 4 sets of 4 making a total of 16 teammates.

Exercise 1: Use the diagram below to solve the following question.

In a pond near Homer's nuclear power plant, there are 3 one-eyed starfish to every 1 porcufish. If there are 30 one-eyed starfish, how many porcufish are there in the pond?

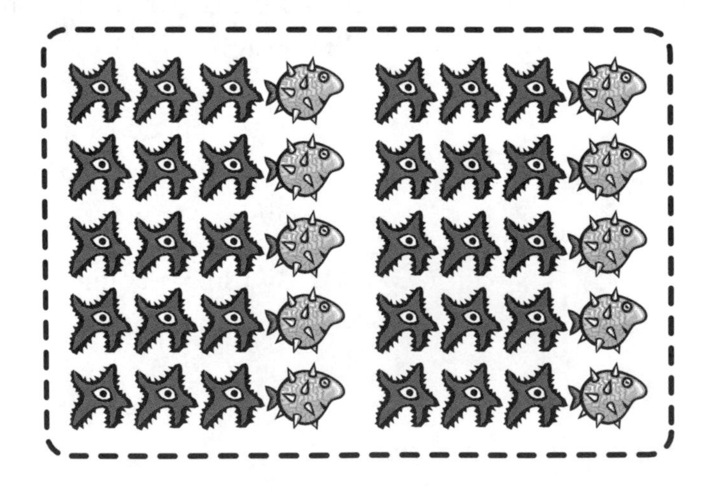

You just planted yourself a garden. You decided to go with a 3:2 ratio of tulips to daisies. You have 36 tulips. How many flowers do you have in all?

Your Garden:

Thus, you have 12 groups of 5, or 60 flowers.

Coincidentally, the day after you plant all of your flowers, your neighbor plants tulips and daisies in the same 3:2 ratio. You nonchalantly stroll past his house and quickly count that he has 14 daisies; how many flowers does he have in all?

You know that when dealing with a 3:2 ratio, the total is going to be a multiple of 3 + 2, or 5. You just do not know how many sets of 5 the neighbor has . . .

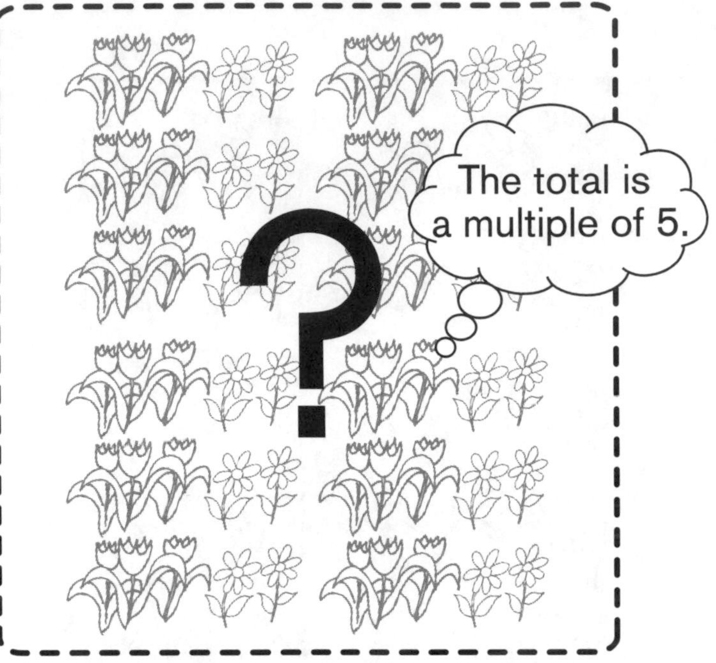

But, you do know that he has 14 daisies . . .

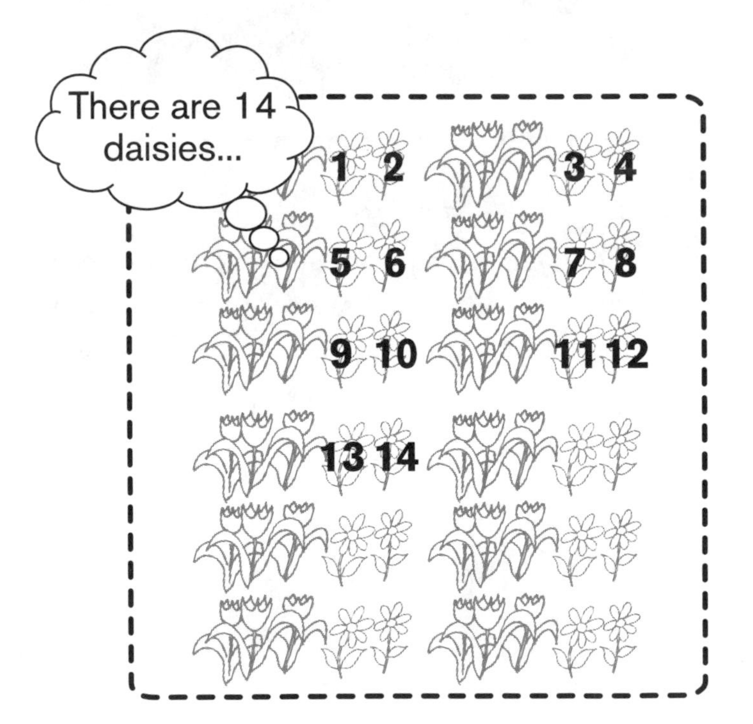

Thus, he must have $5 \times 7 = 35$ flowers in all.

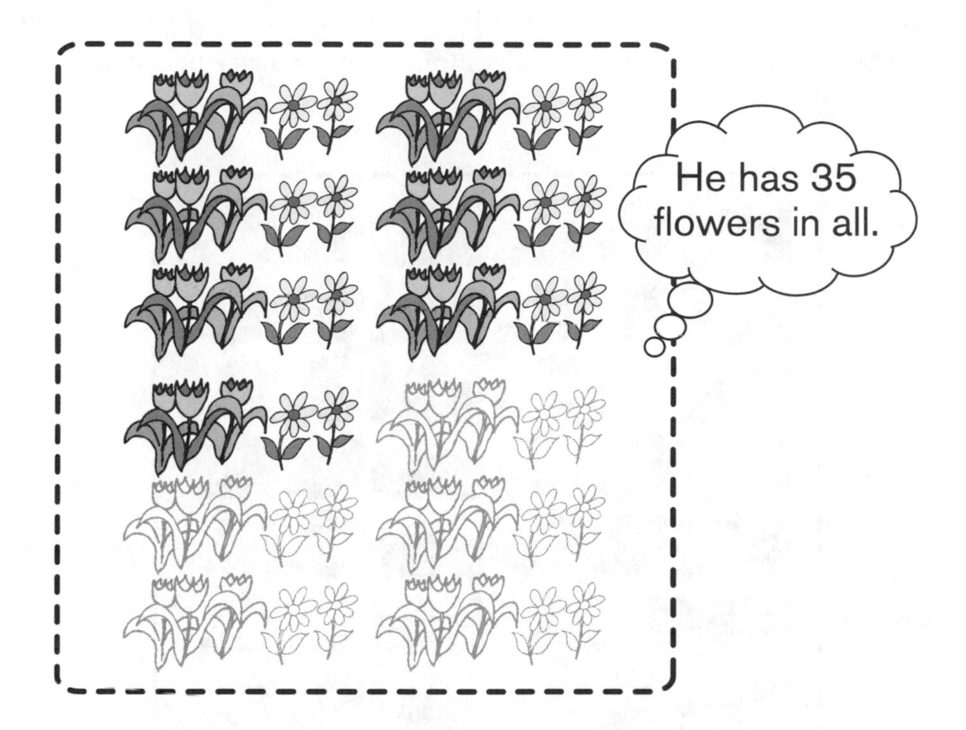

Exercise 2: Use the diagram below to solve the following question.

In an art classroom, there is a ratio of 7 pencils to 2 rolls of tape. If there are 108 pencils and rolls of tape combined, how many pencils are there?

Now you'll look at a trickier ratio question. This question involves a ratio that changes. There are currently 24 people at a costume party and there is a 2:1:1 ratio of Elvis impersonators to Napoleons to Caesars.

I wish I was at this party.

How many consecutive Napoleons must walk through the door in order for there to be a 2:2:1 ratio of Elvis impersonators to Napoleons to Caesars?

The current 2:2:1 ratio means that the total is a multiple of 2 + 1 + 1 = 4. You know there are 24 people present, so there must be 6 groups of 4:

We need to squeeze a new Napoleon into each set in order to generate the desired 2:2:1 ratio . . .

Thus, 6 Napoleons must walk through the door.

Let's look back at your team. As we reminisce, let's contemplate the algebra behind solving ratio questions. Let's look at the team again:

Everybody's favorite variable is x, so let's take that 3:1 boy to girl ratio and say the number of boys is 3 times some number, or $3x$. The number of girls is 1 times the same number, or $1x$, and the total is then $4x$. Here you can just look at the team picture to see that x equals 4 because there are 4 groups of 4. There are 3×4, or 12 boys. There are 1×4, or 4 girls. And there are 4×4, or 16 players in all.

When you looked at the garden with the 3:2 ratio of tulips to daisies, you were given a situation in which you knew that the neighbor had 14 daisies and you wanted to figure out the total.

You knew the total for a 3:2 ratio would be a multiple of 3 + 2, or 5. Using an x you could write:

$$tulips + daisies = total$$

$$3x + 2x = 5x$$

When you were given that there were 14 daisies, you were actually given the value of $2x$.

$$tulips + \boxed{daisies} = \text{total}$$

$$3x + \boxed{2x} = 5x$$

Knowing $2x = 14$, means $x = 7$. If x is 7, then the total, or $5x = 5 \times 7 = 35$.

Algebra comes in handy when working with large numbers . . .

At Maynard High School there used to be a 2:3 ratio of males to females. After a group of males enrolled, the ratio changed to 4:3. If there are 350 students in the school now, how many people were in the school before the new males enrolled?

To solve a question like this, you need to look at the before and after. *Before* you had a 2:3 ratio. The total was a multiple of 2 + 3, or 5. You do not know how many groups there were, but each group looked like this:

Before:

The males will be $2x$ and the females will be $3x$, so that the 2:3 ratio is preserved.

$$\textit{males} + \textit{females} = \textit{total}$$

$$2x + 3x = 5x$$

After, you have a 4:3 ratio, and the total is a multiple of 4 + 3, or 7. Since there are 350 students, you must have 350 ÷ 7 = 50 groups that look like this:

If there are 50 groups that look like that, there must be 4 × 50 = 200 males. Also, there must be 3 × 50 = 150 females. Because you know that the number of females remained constant, you know that the $3x$ in the ***before*** scenario must equal 150.

$$\textit{males} + \textit{females} = \textit{total}$$

$$2x + 3x = 5x$$

If $3x = 150$, then $x = 50$. This means that the ***before*** total was $5x = 5 \times 50 = 250$.

proportions

Proportions are just two equal ratios. When written as $a{:}b = c{:}d$, we call a and d the **extremes** (they are on the **e**nd), and we call b and c the **means** (they are in the **mi**ddle).

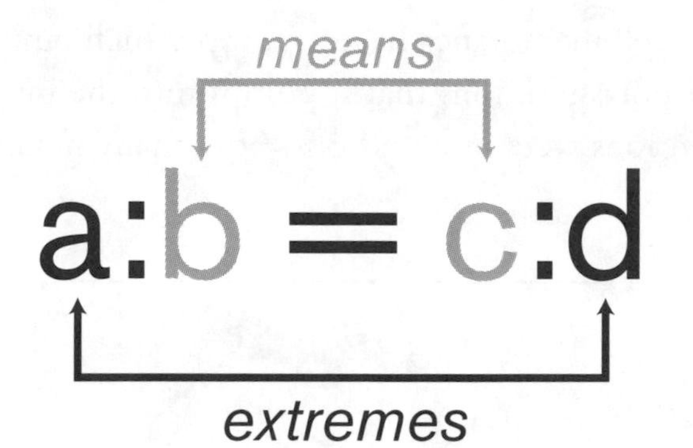

Note that the product of the means equals the product of the extremes. We usually set proportion up by using ratios with bars. You can then cross multiply to solve for an unknown.

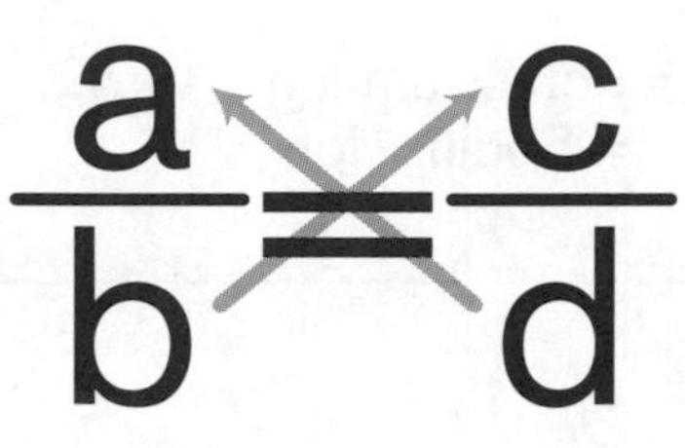

When you set up a proportion, just be sure to align the units correctly. For example, suppose that a cat that is 1 foot tall casts a shadow that is 2 feet long. How tall is a lamppost that casts an 18-foot shadow?

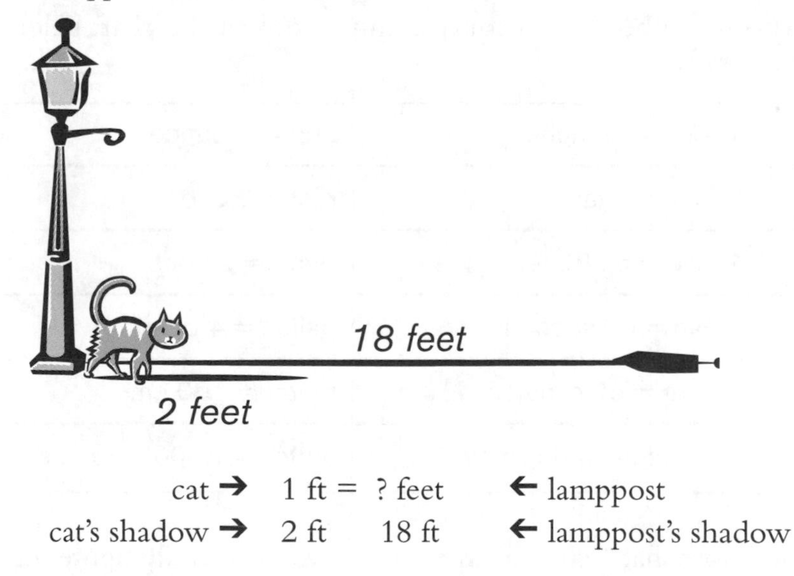

cat →	1 ft =	? feet	← lamppost
cat's shadow →	2 ft	18 ft	← lamppost's shadow

Cross multiply to get 18 × 1 = 2 × *?*, or 18 = 2 × *?*. Dividing both sides by 2, you get *?* = 9 feet.

Exercise 3: Many cell membranes have a pump which pumps 3 sodium ions out for every 2 potassium ions that it pumps into the interior of the cell. If 2,400 sodium ions were pumped out, how many potassium ions were pumped in?

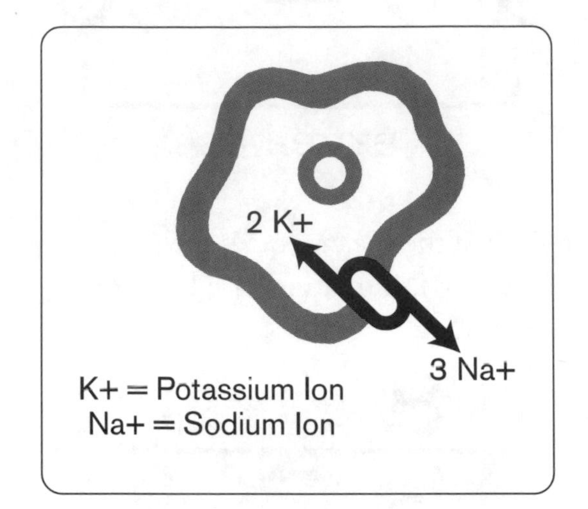

converting units

Proportions can be used to convert units. Look at the chart below:

1 foot = 12 inches	1 cup = 8 ounces
3 feet = 1 yard	1 pint = 2 cups
1 mile = 5,280 feet	1 quart = 2 pints
1 min = 60 seconds	1 gallon = 4 quarts
1 hour = 60 minutes	1 meter = 100 cm
1 meter = 10 decimeters	1 meter = 1,000 millimeters

If you know that 1 cup equals 8 ounces, you can easily figure out how many cups are in 64 ounces by setting up a proportion:

$$\frac{1 \text{ cup}}{8 \text{ oz}} = \frac{? \text{ cup}}{64 \text{ oz}}$$

Cross-multiplying, you get 64 • 1 = 8 • ?, or 64 = 8 • ?. Dividing both sides by 8, you get ? = 8.

conversion factors

Conversion factors are just a "version of 1" because the top and bottom are equivalent. For example, $\frac{1\text{ ft}}{12\text{ in}}$ is a conversion factor. You can multiply a quantity that you have by a conversion factor whenever you want to get rid of unwanted units. Why would you bother with conversion factors when you can just set up proportions? Consider this . . .

A snail slides along at a rate of 2 cm per minute. How many meters does it move in an hour?

Here you have to convert centimeters to meters and minutes to hours. You can do this easily with conversion factors:

Unwanted units on top...

$$\frac{2\text{ cm}}{\text{min}} \times \frac{1\text{ m}}{100\text{ cm}} \times \frac{60\text{ min}}{1\text{ hr}}$$

...so put unwanted units on bottom

Notice that you strategically placed the centimeters in the bottom of the conversion factor so that it would cancel out the unwanted units on top.

To get rid of the "unwanted" minutes on the bottom, you make a conversion factor that has minutes on top:

Unwanted units on bottom...

$$\frac{2\ cm}{min} \times \frac{1\ m}{100\ cm} \times \frac{60\ min}{1\ hr}$$

...so put unwanted units on top

You multiply to get $\frac{120\ m}{100\ hr} = \frac{1.2\ m}{hr}$.

Exercise 4: Convert $\frac{5\ m}{min}$ into $\frac{cm}{sec}$ by using conversion factors:

Convert units prior to setting up proportions!

You should always convert your units *before* you set up a proportion. Suppose you knew that 8 inches of ribbon costs $.60, and you needed to order 20 feet of it. First, you would convert the 20 feet into inches. You know 1 ft = 12 in, so you multiply:

$$\frac{20\ \cancel{ft} \times 12\ in}{1\ \cancel{ft}} = 240\ inches$$

Now you know that you need a price for 240 inches. You set up a proportion:

$$\frac{8\ inches}{\$.60} = \frac{240\ inches}{\$\ ?}$$

$$8''\ (\$\ ?) = (\$.60)(240'')$$

$$8''\ (\$\ ?) = 144''$$

$$\$\ ? = \$18$$

direct proportions

When two quantities are directly proportional, it just means that if one quantity increases by a certain factor, the other quantity increases by the same factor. If one decreases by a certain factor, the other decreases by that same factor.

For example, the amount of money you earn may be directly proportional to the amount of hours that you put in. This is true if you are in a situation where, if you work twice as long, you make twice as much. If you work $\frac{1}{2}$ the amount you usually work in one week, you earn $\frac{1}{2}$ the amount you usually make.

In the equation **Force = mass × acceleration** (otherwise known as **F = ma**), you can see that for a given mass (like a manned car) the force of impact is *directly proportional* to the acceleration. Suppose a car hits a tree while traveling at a certain acceleration . . .

How much greater would the force of impact be if the car was accelerating twice as fast?

Well, you know that the mass is going to remain the same, so "lock" that number in the formula. In order for both sides to be equal, if you multiply one side of the equation by 2 (by doubling the acceleration), then you must also multiply the other side of the equation by 2 (hence, the force doubles as well) . . .

Exercise 5: Which graph do you think could represent the relationship between dollars earned versus hours worked? Which one would represent the relation between force and acceleration?

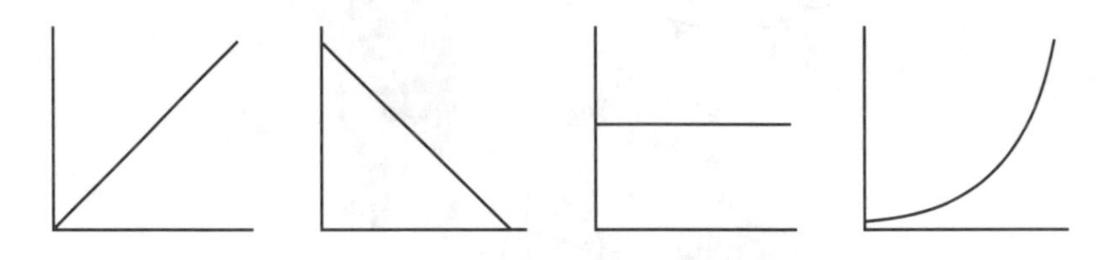

inverse proportions

Two quantities are inversely proportional if an increase by a certain factor for one is accompanied by a decrease by that same factor for the other. We find a lot of these types of relationships in science, so let's take a look at some physics.

Once upon a time, physicists dreamed of a land where the flow of fluids was *ideal*—that is to say that fluids weren't subjected to the perils of resistance, turbulence, and the like. These physicists loved this idea so much that they came up with a formula for *ideal flow*. The ideal flow, Q, which is a constant, is equal to the product of the cross-sectional area times the velocity:

$$Q = A \cdot V$$

Now it's your turn to imagine flowing fluids. Imagine that water is flowing through a pipe and suddenly the pipe widens to a cross-sectional area that is twice as big:

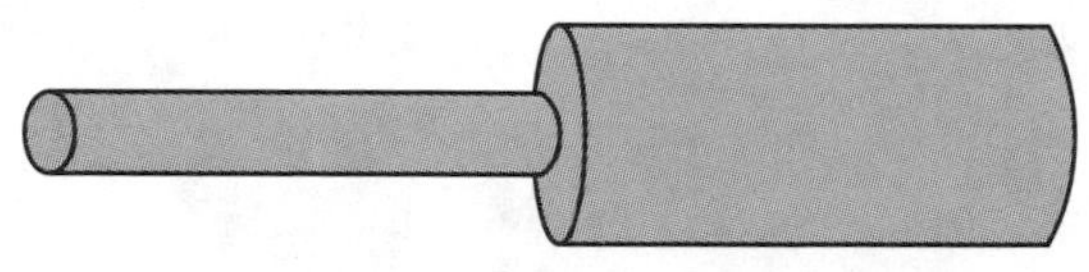

What happens to the velocity of the water?

If you increase A, then in order for Q to remain constant, what happens to V? *Right!* It decreases.

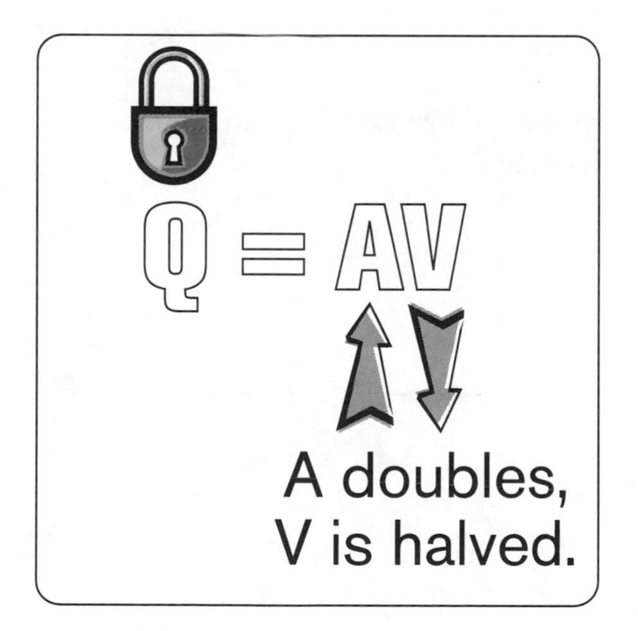

In fact, A gets multiplied by 2, V gets multiplied by $\frac{1}{2}$, and Q remains the same.

$$Q = (A \cdot 2)\ (V \cdot \tfrac{1}{2})$$

$$Q = AV \cdot 2 \cdot \tfrac{1}{2}$$

$$Q = AV \cdot 1$$

$$Q = AV$$

Q is unchanged.

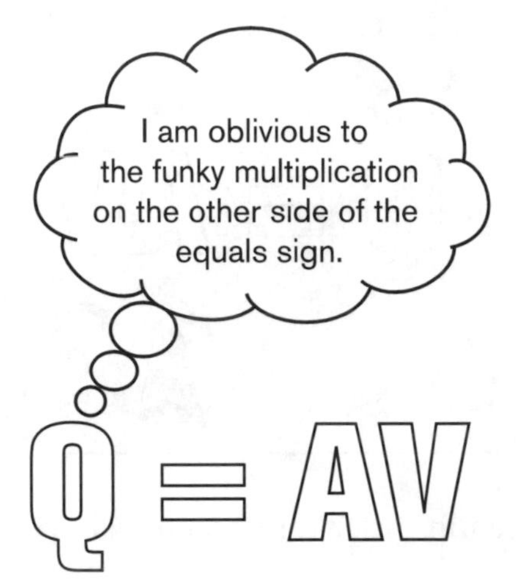

Exercise 6: Which graph do you think could represent the relationship of A and V as described above?

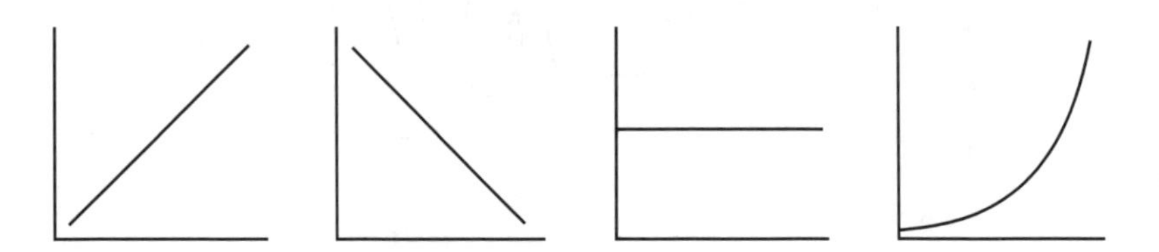

solutions to chapter exercises

Exercise 1: Use the diagram to count the 30 one-eyed starfish. For each set of 3 starfish, there is 1 porcufish, so there must be 10 porcufish.

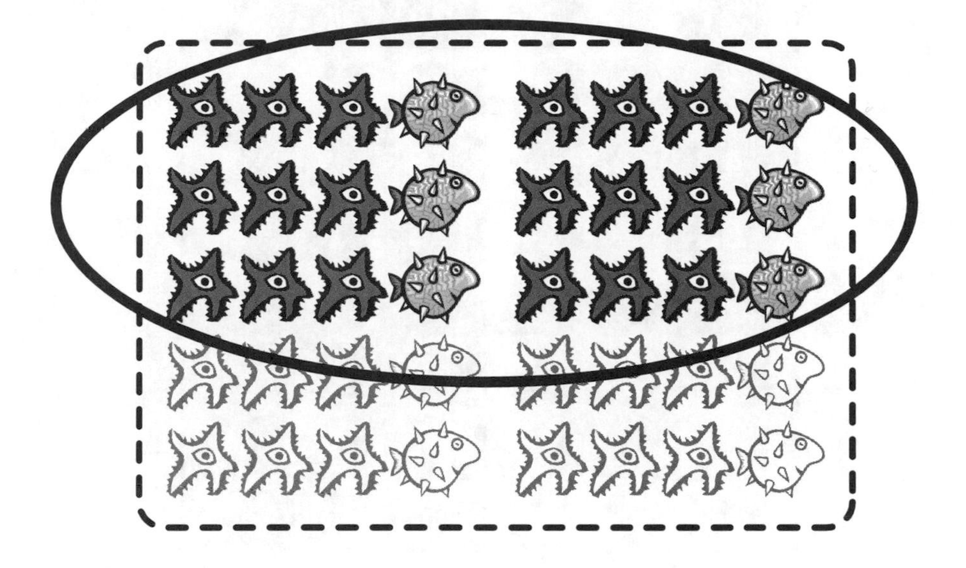

Exercise 2: There is a 7:2 ratio, so the total is a multiple of 7 + 2, or 9. As a matter of fact, you were given a total of 108. This means there are 12 sets of 9. If you count up all of the pencils in these 12 sets, you get . . .

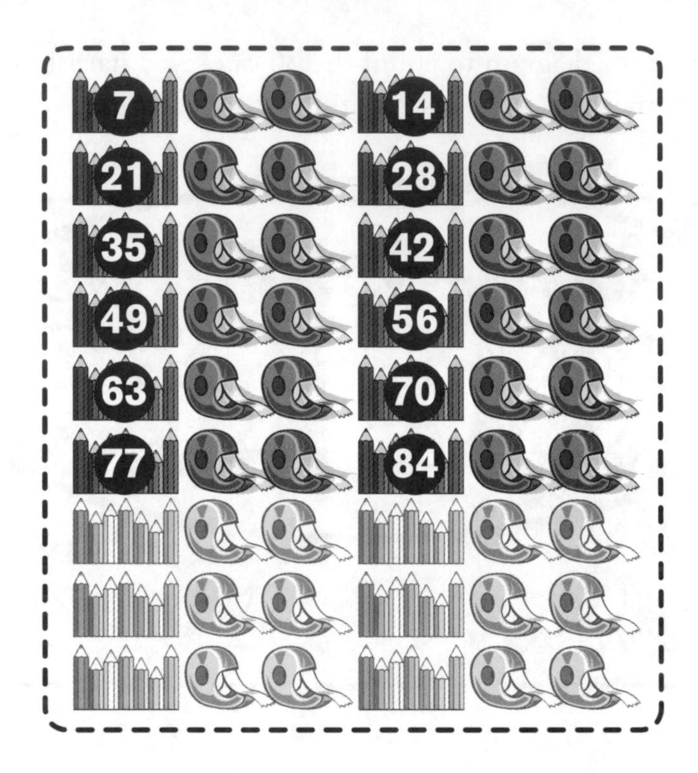

Exercise 3: The pump pumps 3 sodium ions out for every 2 potassium ions that it pumps in. If you want to find how many potassium ions were pumped in when 2,400 sodium ions were pumped out, you just set up a proportion:

$$\frac{3 \text{ sodiums}}{2 \text{ potassiums}} = \frac{2{,}400 \text{ sodiums}}{? \text{ potassiums}}$$

Cross multiplying, you get 3 • ? = 2 • 2,400, or 3 • ? = 4,800. Dividing both sides by 3, you get ? = 1,600. Thus, 1,600 potassium ions were pumped in.

Exercise 4: Using the facts 1 m = 100 cm and 1 min = 60 seconds, you can generate two conversion factors. Be sure to arrange your units so that you can cancel out ones you want to get rid of.

You can get rid of meters by putting meters on the bottom of your conversion factor . . .

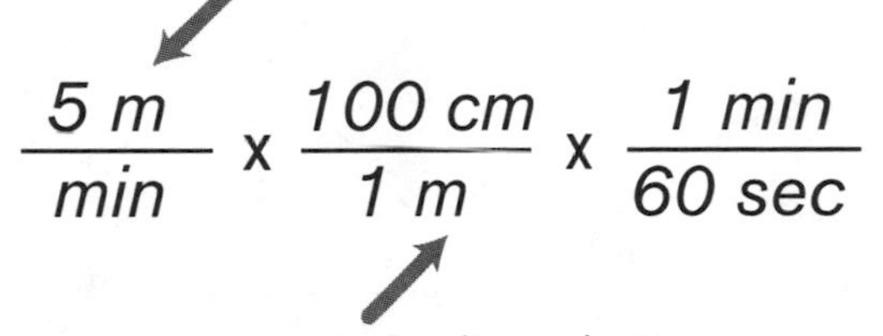

You can get rid of minutes by putting minutes on top in our conversion factor . . .

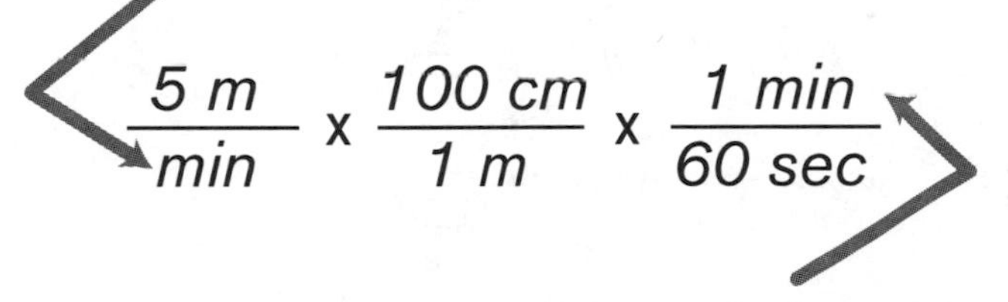

You multiply to get $\frac{500 \text{ cm}}{60 \text{ sec}} = \frac{8\frac{1}{3} \text{ cm}}{\text{sec}}$.

Exercise 5: Dollars earned versus hours worked . . .

Force versus acceleration . . .

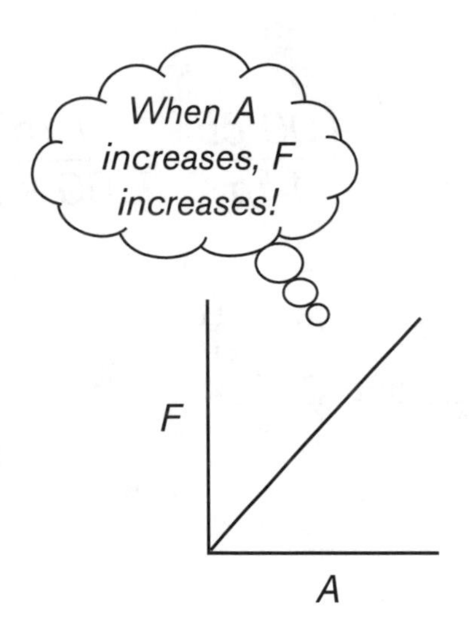

Exercise 6: You saw that as A increases, V decreases . . .

chapter
four

Percents

what is a percent?

Percents are a way of creating a special ratio. When you see a number followed by a percent symbol—%, you just write a ratio comparing that number to 100. For example, $30\% = \frac{30}{100}$.

You can express a percent in two ways: as a fraction (just put the number over 100), or as a decimal (move the decimal point two places to the left). These two options are summarized on the following page:

For example:

$$40\% = \frac{40}{100} = .40$$

Let's look at 25%. You stick 25 over 100 to get $\frac{25}{100}$. Notice that $\frac{25}{100}$ reduces to $\frac{1}{4}$. It is good to be familiar with some on the common fraction and decimal equivalents to percentages. Some are listed in the chart below:

Percent	Fraction	Decimal
25%	$\frac{1}{4}$	.25
50%	$\frac{1}{2}$	.50
75%	$\frac{1}{4}$	.75

Exercise 1: Fill in the chart below. More than one response may be correct.

When I see . . .	I will write . . .
25%	$\frac{1}{4}$
32%	
80%	
100%	
150%	
500%	

taking the percent of a number

Recall that "**of**" means "**multiply**." When you take the percent of a number, you *multiply*.

Let's say you are purchasing an item that usually costs $8, but is now on sale for 25% off. How much do you take off of the $8?

You need to find 25% of $8? Remember that $25\% = \frac{25}{100}$, or $\frac{1}{4}$. So you are taking off $\frac{1}{4}$ of $8.

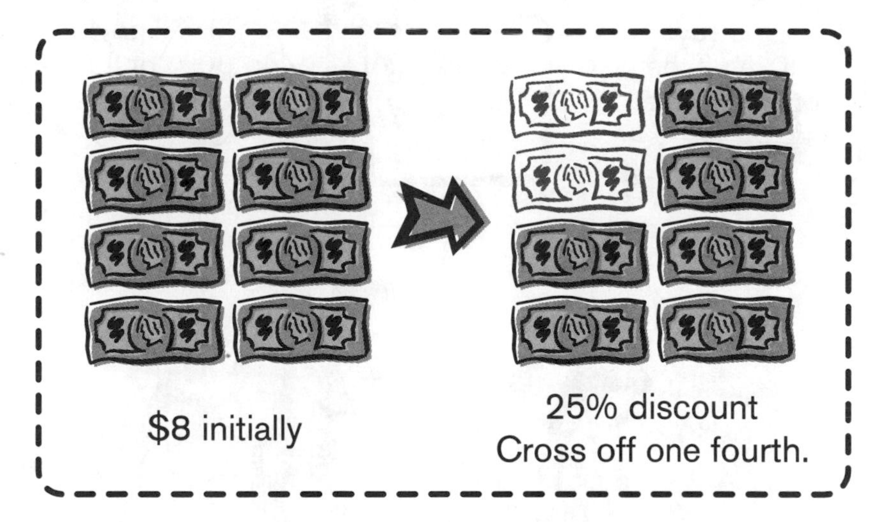

$\frac{1}{4}$ **of** $8 means $\frac{1}{4} \times 8 = \$2$ off.

Notice that you are **saving 25%** of the original price, and you are **paying 75%** of the original price. There are always two ways of looking at the situation:

In the diagram below, you see that a store is having a 20% off sale. Two shoppers are thinking about what this sign means to them:

They are both looking at the situation differently, and they are both correct. Let's say both women find a dress for $80. They want to do a quick mental calculation before deciding whether to buy it or not. Who's going to make a decision quicker? Here is what they each think:

First, the lady with the dog . . .

And, now for the lady with the bags . . .

The lady with the bags made a quicker calculation that did not require subtraction. Eliminating the need to subtract saves time when you are performing mental calculations.

Since we were discussing sales, you were thinking in terms of **"I will save"** versus **"I will pay."** You can use this line of thinking in other situations as well. For example, if a container is 25% (or $\frac{1}{4}$) full, then it is 75% (or $\frac{3}{4}$) empty.

Exercise 2: Evelyn found a dress she loves, selling for 40% off the original $85 price. If she lives in a state that doesn't charge tax on clothes, how much is the dress?

Exercise 3: Jim would like to use the image below for one of his graphics projects. He decides that he wants to reduce the length of the image by 25%, and keep the width the same. Shade in the area that represents how big the image will be after he reduces the length.

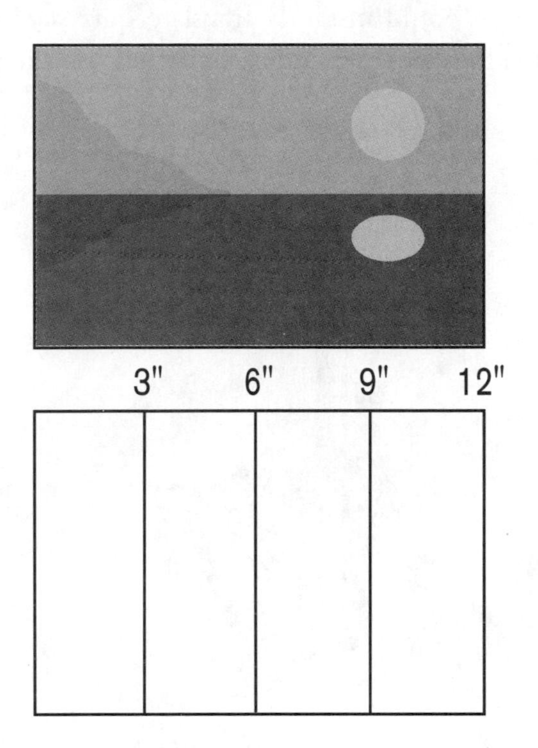

percent of a percent

When you take the percent of a percent, you just multiply. For example if you wanted to know what 40% of 20% of 600 was equal to, you would multiply:

40% of 20% of 600

.40 of .20 of 600

$.40 \times .20 \times 600$

$= 48$

percent proportion

Imagine you had a special **Percent Thermometer** that you could carry around with you that could instantly translate quiz scores to the equivalent grades out of 100? This device would instantly tell you the percentage that you got right:

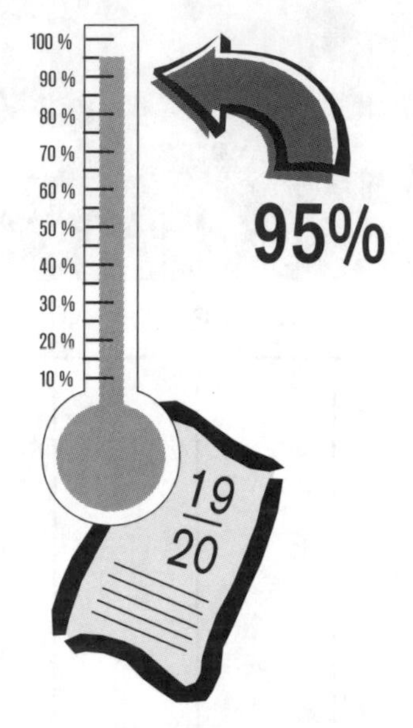

Well, there's no magic thermometer that calculates percentages, but there's an easy way to figure this out for yourself. You just set up a **percent proportion**:

Your score out of 20 ➔ $\frac{19}{20} = \frac{?}{100}$ ← Your score out of 100

Cross-multiply to get $19 \times 100 = 20 \times ?$, or $1{,}900 = 20 \times ?$. Dividing both sides by 20 you get 95. Thus, your score is equivalent to $\frac{95}{100}$, or 95%.

unknown percent

When you see the phrase "four percent," how do you express this mathematically? That's right, you write $\frac{4}{100}$.

When you see the phrase "twenty-three percent," how do you express this mathematically? You simply write, $\frac{23}{100}$.

When you see the phrase "what percent," how do you express this mathematically? Hmmmm. Some of you are thinking, "***I don't know***." Well, you're right:

$$\text{"What percent" means: } \frac{\textbf{I don't know}}{\textbf{100}}$$

Well, for the most part, we mathematicians don't go around writing, "I don't know" in the place of the numbers that we don't know. We are partial to letting ***x*** represent the "unknowns" that we come across.

$$\text{"What percent" means: } \frac{x}{100}$$

Here's an example: What percent of 250 is 30? Let's break this down:

"What percent" means: $\frac{x}{100}$

"of 250" means: $\bullet\ 250$

"is 30" means: $= 30$

So you have:

$$\frac{x}{100} \bullet 250 = 30$$

$$\frac{x \bullet 250}{100} = 30$$

You cross-multiply to get:

$$x \bullet 250 = 30 \bullet 100$$

$$x \bullet 250 = 3{,}000$$

Dividing both sides by 250, you have:

$$x = 12$$

Exercise 4: What percent of 60 is 2? Break it down:

"What percent" means:

"of 60" means:

"is 2" means:

Put it all together and solve:

percent change, percent increase, and percent decrease

Here's a secret. If you ever see a question involving **percent change**, **percent increase**, or **percent decrease**, you can use this multi-purpose formula. That's right! No more stressing over terminology, folks. Too many people have been ripping their hair out over percent increase, percent decrease, and percent change! The madness must stop! Memorize this formula, and you can tackle all three types of questions as easily as a stampeding bull tackles the man with the red scarf.

Use this formula for:

- Percent Increase
- Percent Decrease
- Percent Change

What you put in this formula:

- The **"change"** is just the **change in value**. If something was $17 and now it is $10, the change is $7.
- The **"initial"** is the **initial value**. If something was $17 and now it is $10, the initial value is $17.

Sticking the **change** over the **initial** creates a ratio of the *change in value* to the *initial value*. (The ratio is $\frac{\text{change}}{\text{initial}}$.) This is the premise of any percent change, percent increase, or percent decrease question. Once you know the ratio of the change to the initial, you can figure out how much of a change out of 100 this is equivalent to. How much of a change is it? *I don't know*, you say. Hence, the "?" in the equation.

Your change out of initial ➔ $\frac{\text{change}}{\text{initial}} = \frac{?}{100}$ ← Your change if out of 100

Let's apply this formula. Jade invested $80 in tech stocks. By the next week, her stock was already worth $120. What is the percent increase in the stock value?

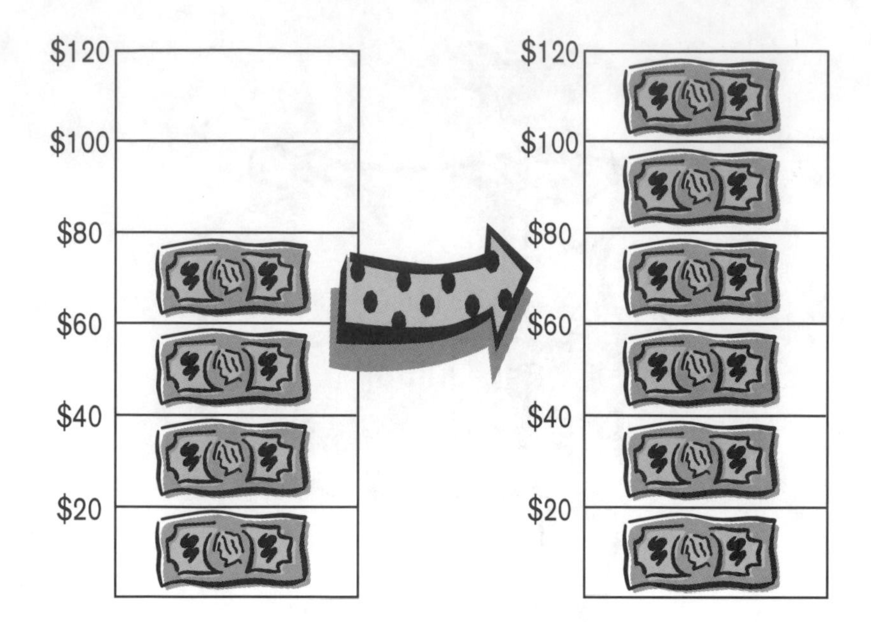

You use:

$$\frac{\text{change}}{\text{initial}} = \frac{?}{100}$$

Here the change is $120 – $80 = $40. The initial value is $80. You put these numbers into the formula: $\frac{40}{80} = \frac{?}{100}$. Cross-multiplying, you get: 4,000 = 80 • ?. You divide both sides by 80 to get ? = 50. Thus, there was a 50% increase.

You can see that this is true:

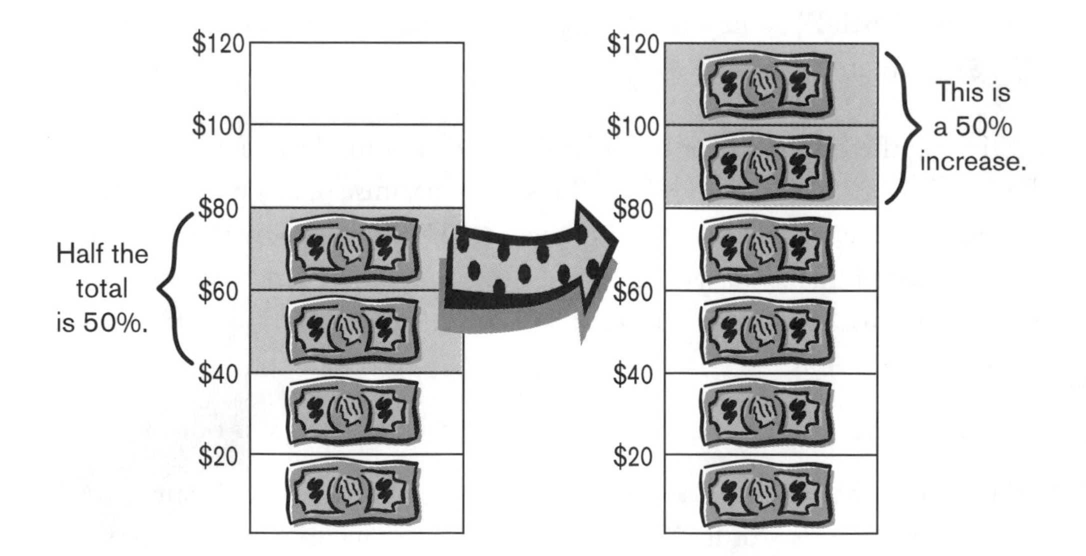

Next, consider a percent decrease. The snowfall in your town was 120 inches throughout the year 2001, and only 90 inches throughout the year 2002. Calculate the percent decrease.

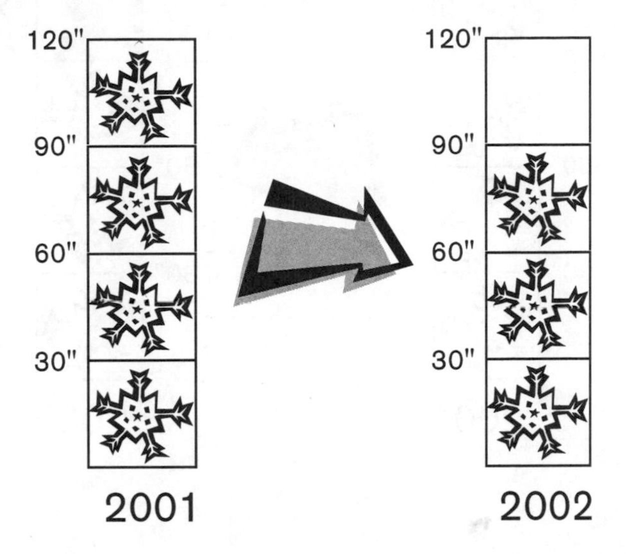

You use:

$$\frac{\text{change}}{\text{initial}} = \frac{?}{100}$$

Here the change is 120 inches – 90 inches = 30 inches. Note that you don't have to worry about the change being –30. You know that the snowfall decreased, so you know you are calculating a percent decrease. Don't even bother with the negative sign.

The initial value is 120 inches. You put these numbers into the formula: $\frac{30}{120} = \frac{?}{100}$. Cross-multiplying, you get: 3,000 = 120 • ?. You divide both sides by 120 to get ? = 25. Thus, there was a 25% decrease.

You can see that this is true:

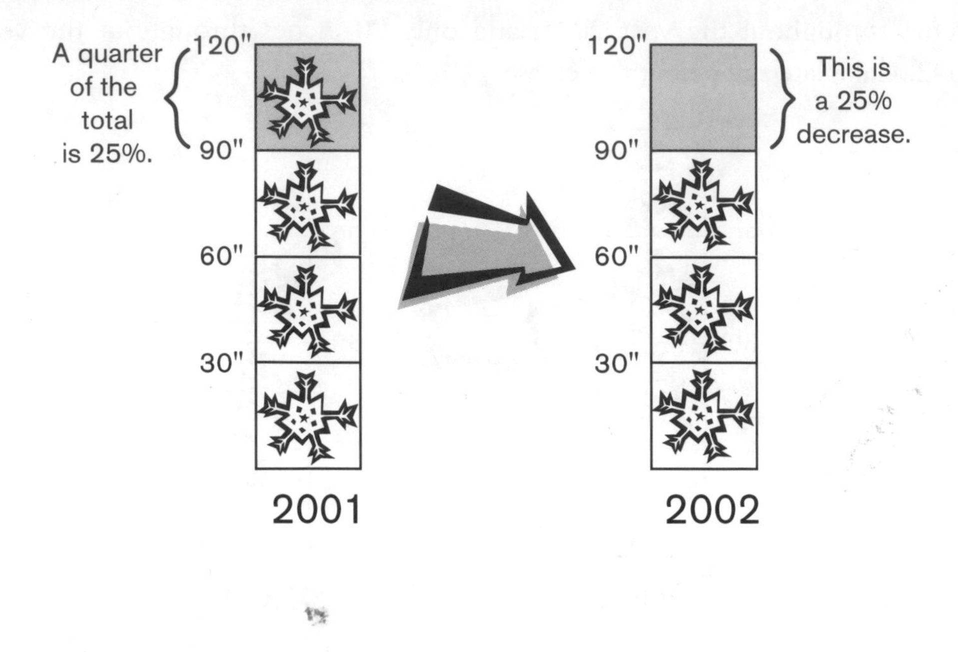

Exercise 5: Refer to the chart below in order to calculate the percent change in recalled trucks from 1998 to 1999.

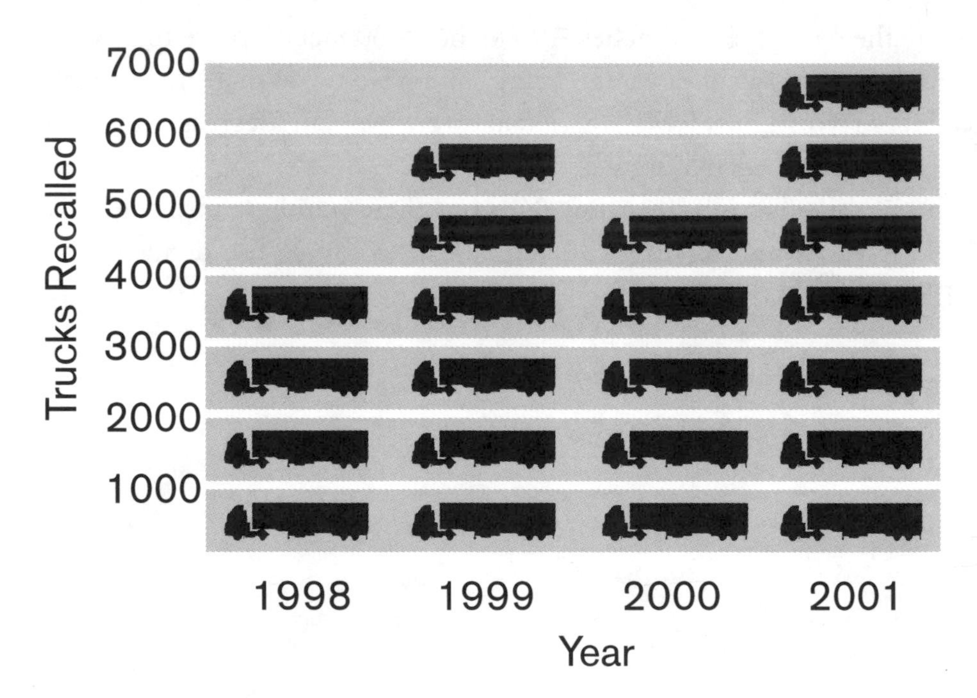

successive percent changes

When you have a situation where there is a percent increase, followed by a percent decrease, followed by another percent decrease, be forewarned:

There are no shortcuts when dealing with successive percent changes!

What you think may be a shortcut may ultimately mean that you get a wrong answer!

Let's look at a 75% decrease followed by a 50% increase. Here, you solve this question the smart way:

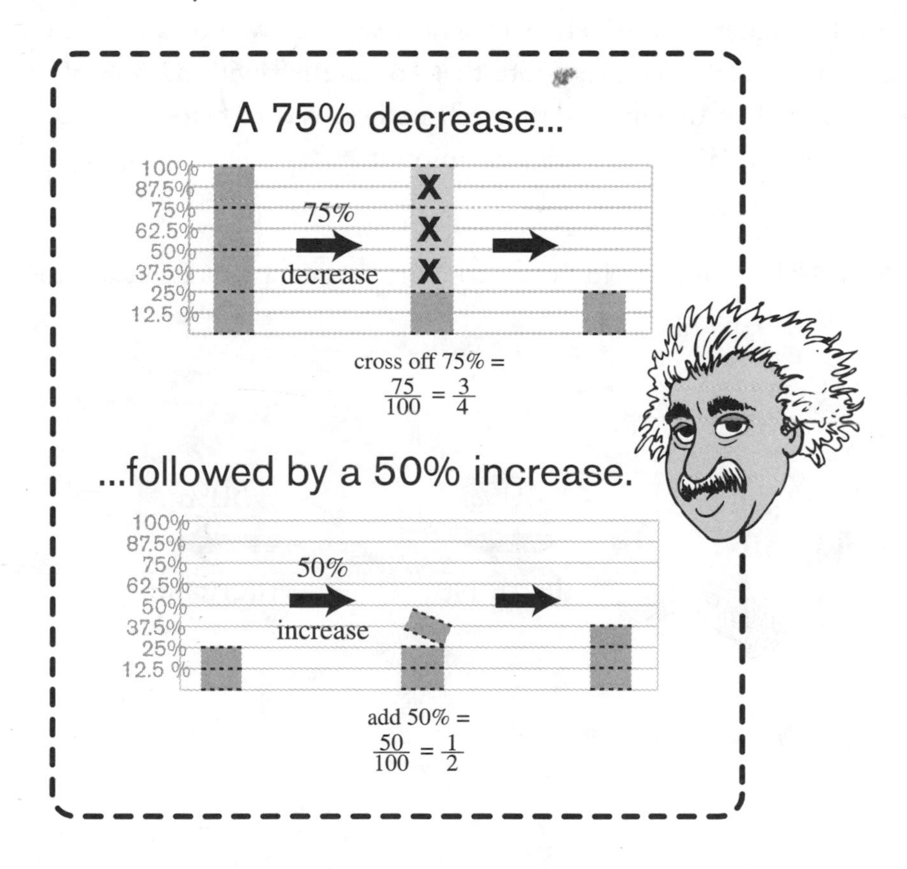

Many see a 75% decrease followed by a 50% increase and think that they could just calculate a 25% decrease. This is **WRONG**:

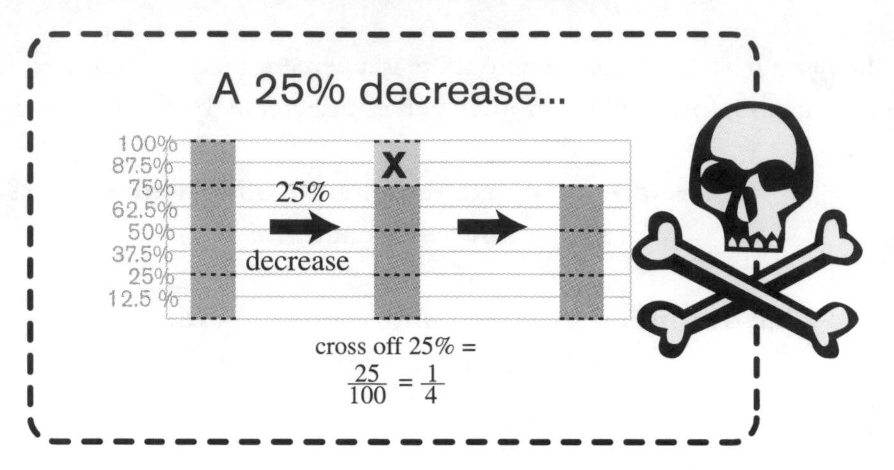

When taken successively, a 75% decrease followed by a 50% increase is the same as a 62.5% decrease (note that you are left with **37.5%** of the original amount). The doomed "shortcut" method was to take a 25% decrease, and that yielded **75%** of the original amount.

Exercise 6: Use the grid below to demonstrate a 40% decrease, followed by a 50% increase.

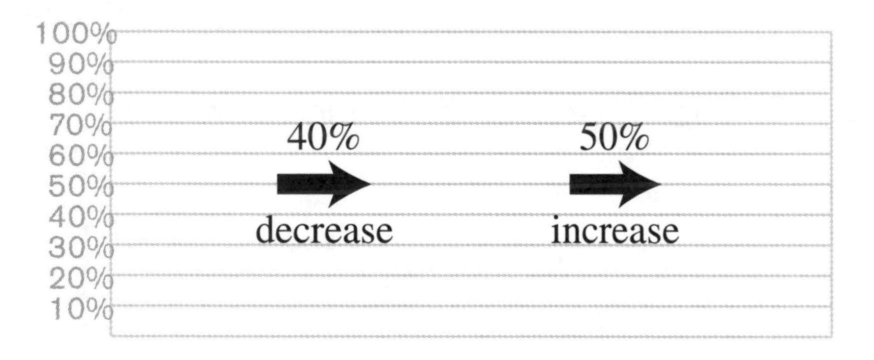

simple interest

You calculate simple interest with a simple formula:

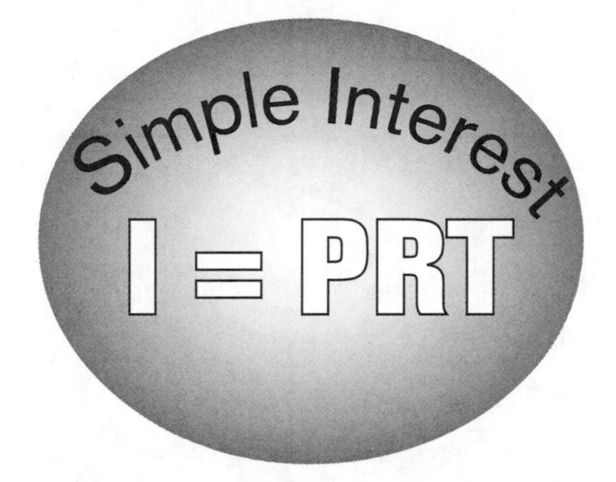

Interest = Principal × Rate × Time

What we put into the I = PRT formula:

- The **interest** is the money you make on top of your initial investment.
- The **principal** is your initial investment. In other words, it is the amount of money you started with.
- The **rate** at which you earn money is expressed as a percent. You may earn money at a rate of 3%. This means you would put R = 3%, or better yet, R = .03 into the I = PRT formula.
- The amount of **time** that your investment earns money is always expressed in years. Make sure you convert the time into years if it isn't given in years.

For example, if the rate of interest is 20% a year, what will the interest be on a $6,000 investment for 2 years?

1. Express the percent as a decimal (or fraction). Thus, R = 20% = .20
2. Make sure that time is in years.
3. Use the formula: I = PRT

$$I = \$6{,}000 \times .20 \times 2$$

$$= \$2{,}400$$

Exercise 7: Zoey earned $200 in 18 months at a rate of 10%. Find the principal.

1. Express the percent as a decimal (or fraction):

2. Make sure the time is in years:

3. Use the formula I = PRT:

compound interest

Sometimes you earn interest on your interest! For example, if you put money into an account that pays 5% interest compounded annually, 5% of your principal is added to your account after the first year. You would then have a new (and larger) principal that earns interest for the second year.

Here is the compound interest formula:

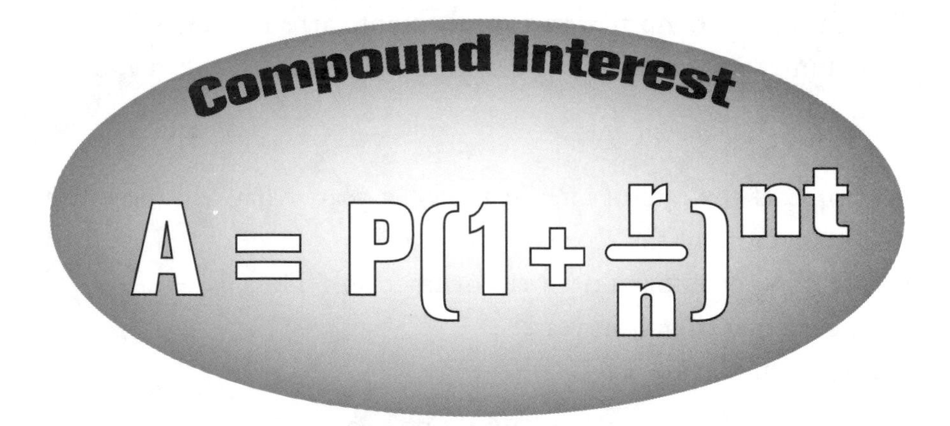

What we put into the $A = P(1 + \frac{r}{n})^{nt}$ formula:

- **A** is the total amount
- **P** is the original principal
- **r** is the rate
- **n** is the number of yearly compounds
- **t** is time (in years)

To find your "**n**" look out for these terms:

- **compounded annually** means interest is paid each year
- **compounded semiannually** means interest is paid two times per year
- **compounded quarterly** means interest is paid four times per year
- **compounded monthly** means interest is paid every month
- **compounded daily** means interest is paid every day

So, let's look at an example. You open a savings account that pays 3% interest semiannually. If you put in $1,000 initially, how much do you have after 2 years?

We use $A = P(1 + \frac{r}{n})^{nt}$, and you substitute in the following values:

$P = 1{,}000$

$r = 3\%$, or $.03$

$n = 2$ (compounded semiannually means twice a year)

$t = 2$

$$A = P(1 + \tfrac{r}{n})^{nt} = 1000(1 + \tfrac{.03}{2})^{2 \cdot 2}$$

$$= 1{,}000(1 + .015)^4$$

$$= 1{,}000(1.015)^4$$

$$= 1{,}000(1.06)$$

$$= 1{,}061.3\overline{6}$$

$$= 1{,}061.37$$

Always round money to the nearest cent. Thus, you'd have $1,061.37.

You ask: *"What if I don't want to memorize that scary formula?"* Well, you have a few options:

- You can do the calculation "the long way." For example you would know that after $\frac{1}{2}$ a year, the \$1,000 principal above would earn I = PRT, or $I = 1{,}000 \times .03 \times \frac{1}{2} = \$15$. Now the account has \$1015. In another $\frac{1}{2}$ year you earn $I = PRT = 1{,}015 \times .03 \times \frac{1}{2} = 15.23$, and you would have \$1,030.23. You would continue calculating in this manner until you completed two years worth of money making.
- You can find out if there is a reference sheet that may contain this formula (if you are taking a standardized test).
- You can use process of elimination on tests. Cross off any preposterous answers and try to pick one that would make sense.
- TIP: In doing a compound interest test question, you know that a lot of people would tend to accidentally solve it as if it were a simple interest question. And you can bet the test designers know this! So, you can cross off the answer that represents I = PRT (the simple interest formula) and pick an answer that is greater.

Exercise 8: Evan opens a savings account that pays 5% interest quarterly. If he put in \$2,000 initially, how much does he have after six months?

algebraic percents

Let's say that Jaclyn buys a printer for *D* dollars and gets a 20% discount. How do you represent this mathematically?

Well, if Jaclyn is getting a 20% discount, she must be paying 80% of the original price. What is the original price? *D.* So she is paying 80% of *D.* This is just $.8 \cdot D$, or $.8D$.

What if she was buying three items that cost *D, E,* and *F* dollars each, and she was getting the same 20% discount on her entire order?

Well, without the discount, her cost would be $(D + E + F)$.

$$\text{non-discounted} = (D + E + F)$$

Now you know that she will only have to pay 80% of the original price, so you multiply the non-discounted price by .8.

$$\text{discounted price} = 80\% \text{ of } (D + E + F)$$

$$\text{discounted price} = .8 \bullet (D + E + F)$$

$$= .8(D+E+F)$$

Brian goes to a restaurant and wants to leave a tip that represents 20% of his bill. If his bill costs B dollars, how much will he spend on the cost of dinner and tip?

Think it through:

- He will pay the bill and leave a tip.
- He will spend $\$B$ on the bill.
- He will pay 20% of B for the tip, or $.2B$ for the tip.
- Bill + Tip = $B + .2B$
- This adds to a $1.2B$ total.

Mental Shortcut:

- He is paying 100% of the bill (the whole bill) plus 20% of the bill, so he is paying 120% of the bill.
- He pays $1.2B$ total.

Consider this scenario: Jane's store sells items for 50% more than she pays for them. If it costs her C dollars to purchase an item, how much will she charge for it?

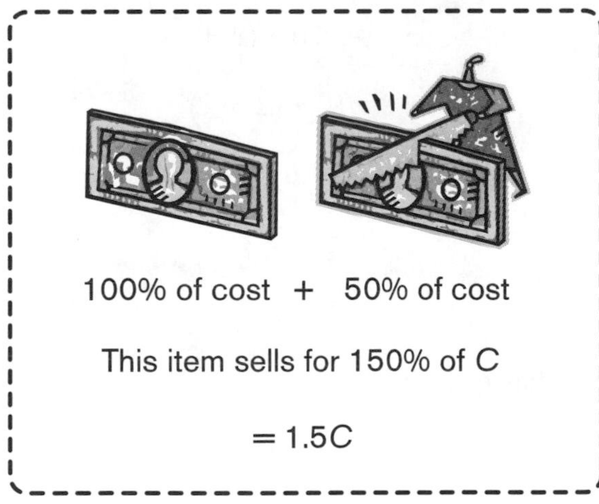

Here's a trickier question: How much water must be added to 2L of a 20% bleach–80% water mixture to yield a 10% bleach–90% water mixture?

Note that you are only adding water to the initial 20% bleach–80% water mixture. This means that the amount of bleach is the same in both mixtures.

The old mixture is 2 liters, and the new mixture is 2 liters plus some unknown amount, or (2 + ?) liters.

You know that the amount of bleach that comprised 20% of the initial 2 liters is the same amount of bleach present when comprising 10% of the new (2 + ?) liter mixture:

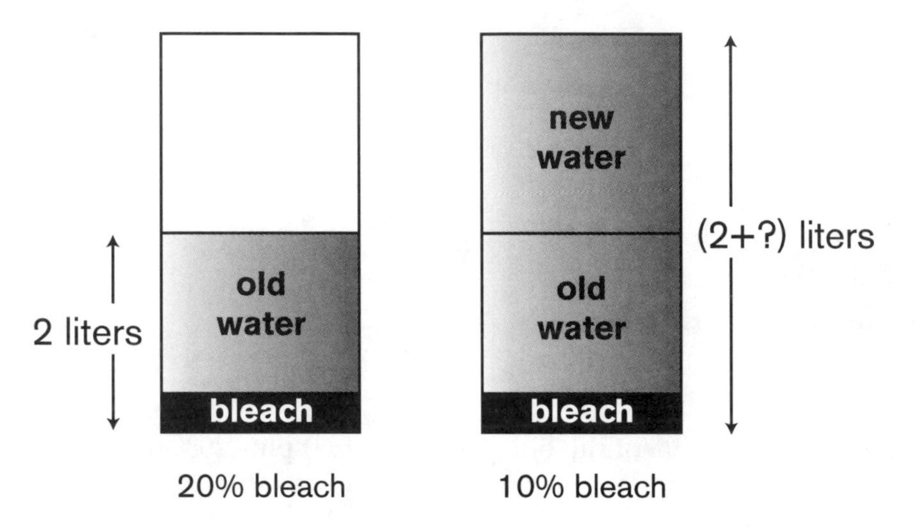

This means that 20% of the 2 liter mixture is the same as 10% of the (2 + ?) liter mixture.

$$20\% \text{ of } 2 \text{ liters} = 10\% \text{ of } (2 + ?) \text{ liters}$$

$$.2 \bullet 2 = .10 \bullet (2 + ?)$$

$$.4 = .10 \bullet (2 + ?)$$

$$\frac{.4}{.10} = 2 + ?$$

$$4 = 2 + ?$$

$$2 = ?$$

Thus, 2L of water has to be added.

solutions to chapter exercises

Exercise 1:

When I see . . .	I will write . . .
25%	$\frac{1}{4}$
32%	$\frac{32}{100}$, $\frac{8}{25}$, .32
80%	$\frac{80}{100}$, $\frac{4}{5}$, .8
100%	$\frac{100}{100}$, 1
150%	$\frac{150}{100}$, $\frac{3}{2}$, 1.5
500%	$\frac{500}{100}$, 5

Exercise 2: If she is **saving 40%** off the price of the dress, then she is **paying 60%** of the price listed. 60% of \$85 can be written mathematically as $.60 \times 85$, which comes out to \$51.

Exercise 3: You want to reduce the length by 25%, or $\frac{1}{4}$.

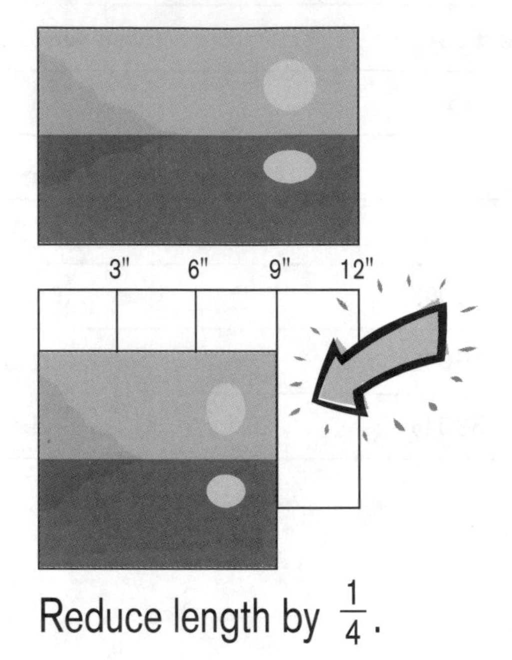

The shaded area below represents the new dimensions of the image:

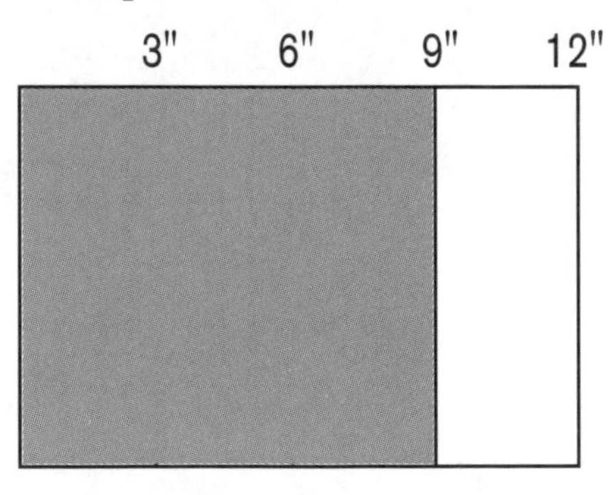

Exercise 4: Given "What percent of 60 is 2?" you break it down as follows:

"What percent" means: $\frac{x}{100}$

"of 60" means: **• 60**

"is 2" means: **= 2**

Put it all together and solve: $\frac{x}{100} \cdot 60 = 2$

$$\frac{60x}{100} = 2$$

$$60x = 200$$

$$x = 3\tfrac{1}{3}\%$$

Exercise 5: In order to calculate the percent change in recalled trucks from 1998 to 1999, use the formula:

$$\frac{\text{change}}{\text{initial}} = \frac{?}{100}$$

Here the change is 6,000 – 4,000 = 2,000.

The initial value is 4,000.

You put these numbers into the formula: $\frac{2{,}000}{4{,}000} = \frac{?}{100}$. You reduce to get $\frac{1}{2} = \frac{?}{100}$. Cross-multiplying, you get: 100 = 2 • ?. Divide both sides by 2 to get ? = 50. Thus, there was a 50% change. Since you went from 4,000 to 6,000, you know that the change was +50%.

You can see that this is true:

This is a +50% change over the 1998 value.

Trucks Recalled

7000
6000
5000
4000
3000
2000
1000

1998 1999 2000 2001

Year

This is 50% (or half) of the '98 total.

Exercise 6: First decrease by 40%. Next, increase by 50%.

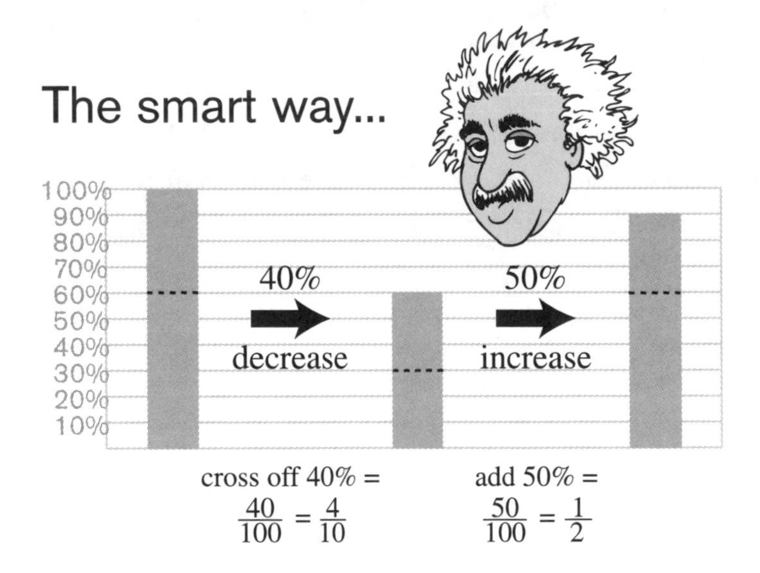

The final result is 90% of the original value.

Exercise 7: You know Zoey earned \$200 (that's the **Interest**) in 18 months at a rate of 10%. Find the principal.

1. Express the percent as a decimal (or fraction):

.1

2. Make sure the time is in years:

18 mo = 1.5 years

3. Use the formula I = PRT

$$I = P \times R \times T$$

$$200 = P \times .1 \times 1.5$$

$$200 = .15P$$

$$P = \$\ 1{,}333.33$$

Exercise 8: Evan opens a savings account that pays 5% interest quarterly. He put in \$2,000 initially, so to find how much he has after 6 months, you use:

$$A = P(1 + \tfrac{r}{n})^{nt}$$

Make sure to convert the 6 months into $\frac{1}{2}$ yr.

$$\begin{aligned} A &= P(1 + \tfrac{r}{n})^{nt} \\ &= 2{,}000(1 + \tfrac{.05}{4})^{4 \cdot \frac{1}{2}} \\ &= 2{,}000(1 + .0125)^2 \\ &= 2{,}000(1.0125)^2 \\ &= 2{,}000(1.0251563) \\ &= 2{,}050.3125 \\ &= 2{,}050.31 \end{aligned}$$

Algebra

variables

Variables are just numbers in disguise. It has always been a traditional pastime for young numbers to pretend that they were secret agents. These numbers tried to conceal their identity, first with sunglasses . . .

. . . but that didn't work out so well. Then they tried using paper bags, but they kept bumping into things. So, finally they decided to take on a new identity altogether! And hence, using a LETTER to represent a NUMBER became the standard in the world of numerical espionage.

Numbers tried really hard to keep their true identities hidden, but cunning sleuths were always able to track down the clues they left behind. The sleuths could then rearrange these clues and deduce the true identity of the secret agents.

Once you realize that these variables are just numbers in disguise, you'll understand that they must obey all the rules of mathematics, just like the numbers that aren't disguised. This can help you figure out what number the variable at hand stands for.

simplifying equations and expressions

Let's look at an expression that doesn't have any variables:

$$3(2) + 4(2)$$

Would this be the same as 7(2)? You quickly solve both expressions and say *Yes!* But what underlying mathematical property tells us that this will be true?

It's the **distributive property**:

$$\mathbf{2(7) =}$$

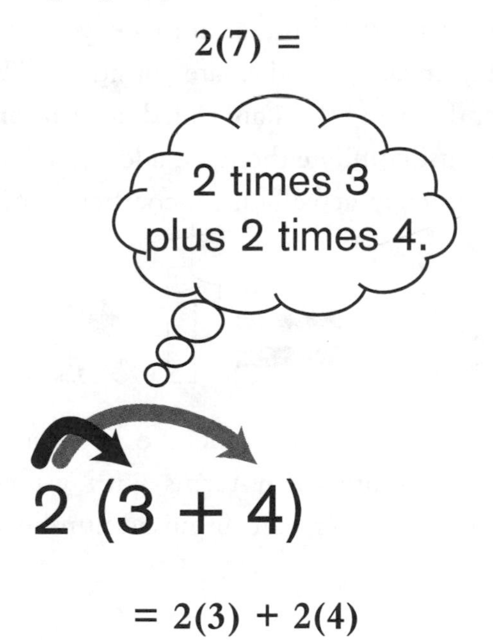

$$\mathbf{= 2(3) + 2(4)}$$

If you saw 2(7), you probably wouldn't change it to 2(3+4) and apply the distributive property, would you? Of course not! But understanding this mathematical concept will help you when you see operations with variables. For example, if you see $3x + 4x$, you can think:

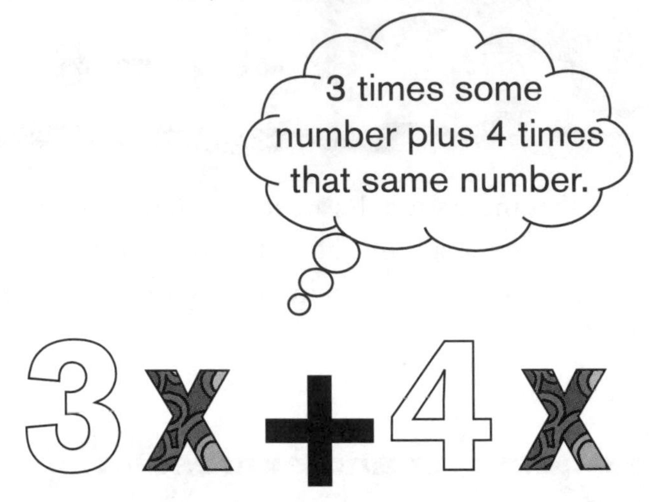

And you would feel secure in saying that $3x + 4x = 7x$; x is really just a number in disguise, and it must obey all the properties that govern numbers.

$$\text{Just as } 3(2) + 4(2) = 7(2),$$

$$3x + 4x = 7x.$$

When you combine the $3x$ and the $4x$ into $7x$, you are "**combining like terms**." What does this mean? $3x$ and $4x$ are considered *like terms* because they both involve x. Here, the 3 and the 4 are called **coefficients** of the x term. If you see $3x + 4x$, you just combine them, or add them to get $7x$.

A term with no coefficient actually has a coefficient of one:

Always combine like terms: Combine the x-terms with x-terms, and x^2-terms with x^2-terms. Combine y-terms with y-terms, combine non-disguised numbers with the other non-disguised numbers, and so forth.

For example, if you see something like $2x + 3x^2 + 4x^2 + 5x$, you can immediately simplify the expression by spotting like terms:

You would then combine like terms to get $7x + 7x^2$.

Exercise 1: Simplify the following expression.

$$2x + 3x^2 + 5x - x^2 + x + 7x^3$$

Sometimes you can simplify an equation by multiplying the entire equation by a certain number. This is mathematically acceptable because you are just generating an equivalent multiple of the equation. Suppose you had $2x = 2$.

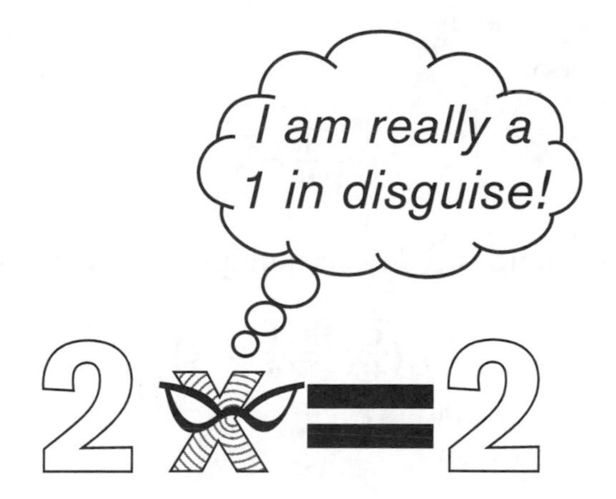

Now let's take a multiple of $2x = 2$. Let's multiply the whole equation by 2:

What if you had multiplied the $2x = 2$ equation by 3?

Let's look at a situation where multiplying the entire equation by a certain number would come in handy.

For example, let's look at:

$$\frac{1}{4}x + \frac{1}{6} = \frac{2}{3}$$

To get rid of the fractions, you can multiply the **entire equation** by 12:

$$\mathbf{12\ (\tfrac{1}{4}x + \tfrac{1}{6} = \tfrac{2}{3})}$$

You distribute the 12 to get:

$$12 \bullet \frac{1}{4}x + 12 \bullet \frac{1}{6} = 12 \bullet \frac{2}{3}$$

$$3x + 2 = 8$$

This is a much simpler equation than the one you started with.

Exercise 2: $(\frac{x}{3}) + (\frac{3x}{10}) - (\frac{2x}{5})$ is equivalent to

a. $\frac{7x}{15}$

b. $\frac{31x}{30}$

c. $\frac{8x}{18}$

d. $\frac{7x}{30}$

solve for *x*

When you want to solve for x, you want to get x all by itself. We call this "**isolating the variable**." In order to preserve the equality of the given equation, you need to be sure that you are doing the same thing to both sides of the equation. This means that you should perform corresponding operations on both sides of the equals sign. If you subtract 2 from the left side, you need to subtract 2 from the right side. If you divide the left side by 3, you must divide the right side by 3.

For example, let's look at $2x + 7 = 15$. You will solve for x.

You have $2x + 7 = 15$. In order to get x by itself, first get rid of the 7. This means you will subtract 7 from both sides.

$$\begin{array}{r} 2x + 7 = 15 \\ -7 \quad -7 \\ \hline 2x = \ \ 8 \end{array}$$

Next, divide both sides by 2 in order to get x by itself:

$$\frac{2x}{2} = \frac{8}{2}$$

$$x = 4$$

Exercise 3: Solve for x.

Given $7x + 2 = 5x + 14$, what is the value of x?

inequalities

Inequalities contain the *greater than, less than, greater than or equal to,* or *less than or equal to* symbols. When you solve the inequalities for x, you can figure out a range of numbers that your unknown is "allowed" to be.

This symbol . . .	Means . . .
$>$	"Greater than"
$\geq$	"Greater than or equal to"
$<$	"Less than"
$\leq$	"Less than or equal to"

Now, there is one rule that you need to remember when dealing with inequalities:

When you multiply or divide by a negative number, you need to reverse the sign.

For example:

$-5x + 3 > 28$ can also be expressed as which of the following?

a. $x < -\frac{31}{5}$
b. $x > -\frac{31}{5}$
c. $x > -5$
d. $x < -5$

This type of question is a lot like the "Solve for x" questions that you did above. The goal here is to isolate your x. First you will subtract 3 from both sides.

$$\begin{array}{r} -5x + 3 > 28 \\ -3 \quad -3 \\ \hline -5x > 25 \end{array}$$

When you multiply or divide by a negative number you need to reverse the sign. So when you divide by -5, you get:

$$\frac{-5x}{-5} > \frac{25}{-5}$$

$x < -5$, which is choice **d**.

On a number line, this answer looks like:

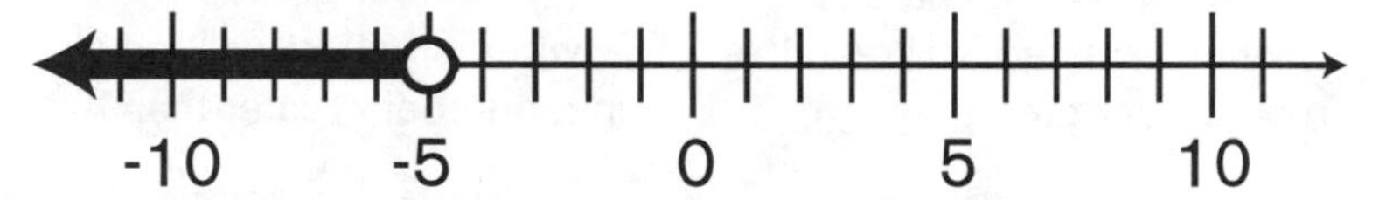

Here's a trick to avoid having to worry about flipping the sign: Just move your terms in a manner such that you will end up with a *positive* coefficient on your variable.

Let's look at $-5x + 3 > 28$ again. Our x coefficient is negative, so you will add $5x$ to both sides.

$$\begin{array}{r} -5x + 3 > 28 \quad \\ +5x \qquad\quad +5x \\ \hline 3 > 28 + 5x \end{array}$$

Next, you subtract 28 from both sides:

$$\begin{array}{l} 3 \qquad\quad > 28 + 5x \\ \quad -28 \qquad\quad -28 \\ \hline 3 - 28 > 5x \\ \quad -25 > 5x \end{array}$$

You divide both sides by 5 to yield:

$$-5 > x$$

Let's look at another example. Suppose you had $x^2 + 12 > 16$ and you wanted to represent this on a number line.

First, try to isolate x. You can get rid of the 12 by subtracting 12 from both sides:

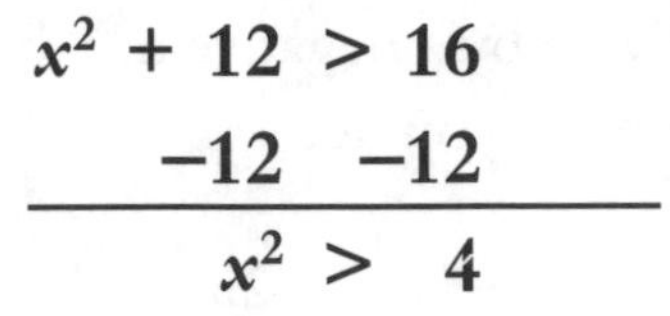

$$\begin{array}{r} x^2 + 12 > 16 \\ -12 \quad -12 \\ \hline x^2 > \quad 4 \end{array}$$

If x^2 is greater than 4, then what do you know about x? What number squared equals 4? Both + 2 and −2, when squared yield 4. Any x greater than 2 will yield an x^2 greater than 4. Any x less than −2 will yield an x^2 greater than 4:

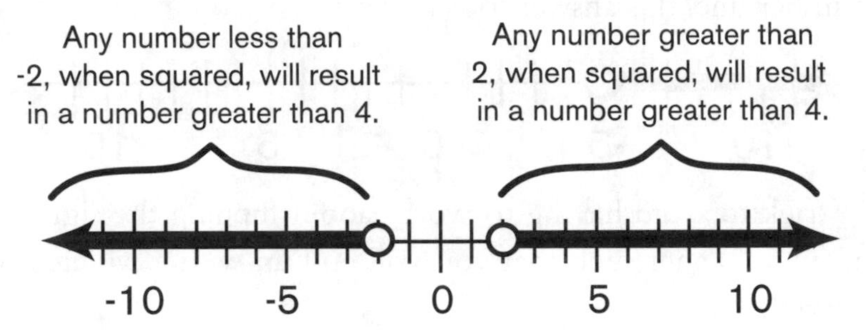

Exercise 4: Graph the solution to $x^2 + 12 \geq 4^2$ on the number line below.

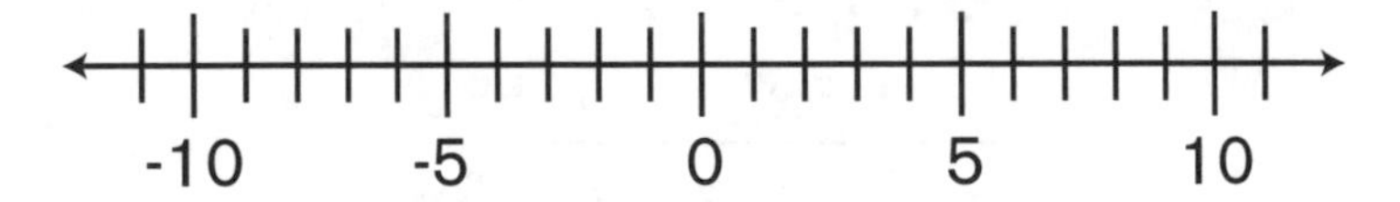

english to equation

In English we can say, "Felicia is 3 years older than Samantha." You can write this mathematically as:

■ Felicia is	$F =$
■ 3 years older than	$+ 3$
■ Samantha	S

You can put this all together: $F = S + 3$.

Let's say Felicia is 3 years younger than Christina. How can you express this mathematically?

■ Felicia is	$F =$
■ 3 years less than	$- 3$
■ Christina	C

You can put this all together: $F = C - 3$. Notice that this means that if you take Christina's age and subtract 3 years, you will end up with Felicia's age.

Let's say that there is 1 more sister. Melissa is twice as old as Samantha. How can you express this?

■ Melissa is	$M =$
■ twice as old as Samantha	$S \times 2$

You put this all together to get $M = S \times 2$, or $M = 2S$.

Exercise 5: How can you represent the following phrase mathematically?

Erik has 3 CDs less than Danny.

a. $D = E - 3$
b. $D = 3 - E$
c. $E = 3 + D$
d. $E = D - 3$

Let's look at a trickier example: Joe only owns 12 more than half the amount of CDs stacked on his dresser. The rest were borrowed from a friend. If there are a total of 52 CDs in the stack, which equation represents the amount of CDs that he borrowed, ***B***?

a. $B = 12 + (\frac{1}{2} \bullet 52)$
b. $B = 52 - 12$
c. $B = \frac{1}{2} \bullet 52 - 12$
d. $B = 52 - (12 + \frac{1}{2} \bullet 52)$

First, realize that there are 52 CDs total, and that some are Joe's and some are the ones he borrowed. So the basic idea would be: **52 total CDs = # Joe's + # Joe borrowed**. You know you should call the borrowed CDs ***B***, and if you similarly call the number of Joe's CDs ***J***, you know $\mathbf{52 = J + B}$. Because you know that you need to find ***B***, rearrange this equation by subtracting ***J*** from both sides:

$$\begin{array}{rcl} 52 & = & J + B \\ -J & & -J \\ \hline 52 - J & = & B \end{array}$$

Hence, you know that $\mathbf{B = 52 - J}$. But none of the answers have a ***J***! This means you need to be more specific about ***J***. What do we know about ***J***, or the number of CDs that Joe owns? Well, the question states that: "*Joe only owns 12 more than half the amount of CDs stacked on his dresser.*" You need to express this statement mathematically. If Joe owns12 more than half the total amount, and you know that the total is 52, then he owns 12 more than $\frac{1}{2}$ of 52. Remember that **more than** means **plus**, and **of** means **multiply**. Mathematically, you know $\mathbf{J = 12 + \frac{1}{2} \bullet 52}$. Now write $\mathbf{12 + \frac{1}{2} \bullet 52}$ in place of ***J*** in the equation $\mathbf{B = 52 - J}$.

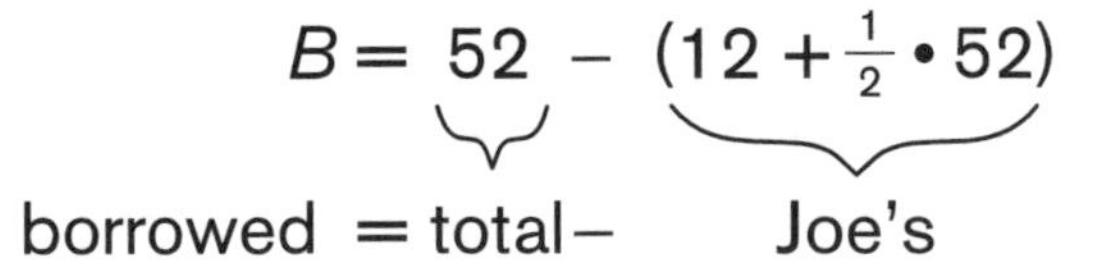

So, the answer is **d.**

substitution

Sometimes algebra is just a matter of sticking, or "substituting," numbers in for the right variables. For example, you are told that $a = b + c$. If $b = 5$ and $c = 7$, what is the value of a? You simply substitute 5 for b and 7 for c to get:

$$a = 5 + 7$$
$$= 12$$

Example: What is the value of $\frac{b^2 + b}{a}$ when $b = 1$ and $a = 2$?

Here you put 1 in for each b you see in the equation. You put 2 in for the a in the denominator:

$$\frac{b^2 + b}{a} = \frac{1^2 + 1}{2} = \frac{1 + 1}{2} = \frac{2}{2} = 1$$

Exercise 6: What is the value of the expression $5x^2 + 2xy^3$ when $x = 3$ and $y = -2$?

a. −3
b. 3
c. −93
d. 93

function tables

Function tables portray a relationship between two variables, such as an x and a y. It is your job to figure out exactly what that relationship is. Let's look at a function table:

x	y
0	–
1	4
2	5
3	–
4	7

Notice that some of the data was left out. Don't worry about that! You can still figure out what you need to do to our x in order to make it our y. You see that $x = 1$ corresponds to $y = 4$; $x = 2$ corresponds to $y = 5$; and $x = 4$ corresponds to $y = 7$. Did you spot the pattern? Our y value is just our x value plus 3.

Exercise 7: The table below shows the relationship between two variables: x and y. Write an equation that demonstrates this relationship.

x	y
0	0
1	–
2	4
3	–
4	16

algebraic formulas

Two algebraic formulas that you should be familiar with are:

- $D = R \times T$
- $w \times d = W \times D$

The first formula is sometimes called the **Constant Rate Equation,** and D = R × T means:

$$\text{Distance} = \text{Rate} \times \text{Time}$$

Distances may be given in miles, meters, feet, etc. Rates may be given in meters per second (m/s), miles per hour (mi/hr or mph), feet per second (ft/sec), and so forth. Just be sure to check that the units you are using will work together. For example, it "works" when you multiply mi/hr by a time in hours, because the hours cross out to yield the distance in miles:

$$\frac{\text{mi}}{\cancel{\text{hr}}} \times \cancel{\text{hr}} = \text{mi}$$

A dog is walking at a rate of 5 meters per minute. 80 meters away, a turtle is walking right at him at a rate of 1 meter a minute. How long before they meet?

It is easy to solve this kind of question if you draw a diagram:

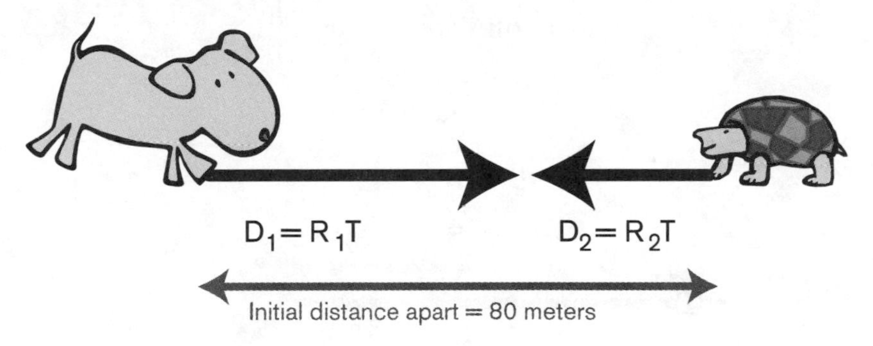

You know that the initial distance apart, 80 m, is equal to $D_1 + D_2$.

$$80 = D_1 + D_2$$

You know $D_1 = R_1 \times T$ and $D_2 = R_2 \times T$, so you can rewrite the above equation as:

$$80 = (R_1 \times T) + (R_2 \times T)$$

You know the dog walks at 5 m/min (this is R_1) and the turtle walks at 1 m/min (this is R_2), so put these two rates into the equation:

$$80 = 5T + 1T$$

$$80 = 6T$$

Dividing both sides by 6 yields $T = 13\frac{1}{3}$ minutes.

One-third of a minute $= \frac{1}{3} \bullet 60$ minutes = 20 seconds. The answer can also be written as 3 minutes and 20 seconds.

Suppose Tamara is standing on the corner with her two pups, and one gets loose. The pup takes off at a rate of $\frac{4\text{ m}}{\text{sec}}$. By the time Tamara notices that her pup broke free, two seconds have passed. She starts running after the dog at a rate of 6 m/sec. How far will she have to run until she catches the dog?

This is a slightly confusing situation, but let's draw a diagram to visualize what is happening. So far we know:

The fugitive pup runs at $\frac{4\text{ m}}{\text{sec}}$.

The pup gets a 2 second head start. Since $D = R \times T$, you know that the pup has a $D = 4 \times 2 = 8$ m head start.

Tamara runs at a rate of $\frac{6\text{ m}}{\text{sec}}$.

So you draw:

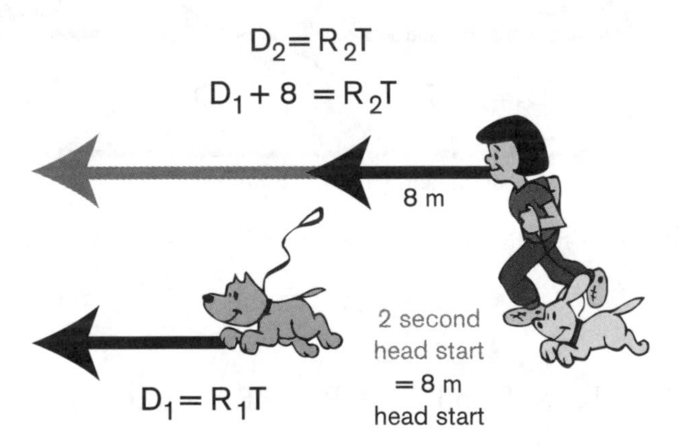

You want to know the amount of time it will take for Tamara to travel the distance labeled D_2. You can see that $D_2 = D_1 + 8$. Use the formula $D_2 = R_2 \times T$, and replace D_2 with $D_1 + 8$.

$$D_2 = R_2 \times T$$

$$D_1 + 8 = R_2 \times T$$

$$4T + 8 = 6T$$

$$8 = 2T$$

$$4 = T$$

It will take her 4 seconds.

Exercise 8: Train 1 is traveling eastbound at 60 mph. On an adjacent track to the east, Train 2 is traveling at 50 mph in a westbound direction. If the two trains departed from their respective stations at the same time and were initially 220 miles apart, how long will it take for them to pass each other?

The second equation that you should know deals with balancing a fulcrum. For a balanced fulcrum, $w \times d = W \times D$ which means:

$$\text{Weight}_1 \times \text{Distance}_1 = \text{Weight}_2 \times \text{Distance}_2$$

Suppose your 180-pound uncle is sitting in his 20-pound easy chair on a balanced fulcrum. He is 4.5 feet from the pivot point. On the other side is an anvil. It has been placed 3 feet away from the pivot point. How much does it weigh?

Let's look at a diagram:

We will call the anvil's weight w, and you will put all the other information into the formula:

$$w \times d = W \times D$$

$$w \times 3 = 180 + 20 \times 4.5$$

$$w \times 3 = 200 \times 4.5$$

$$3w = 900$$

$$w = 300$$

Thus, the anvil weighs 300 pounds.

This equation works for a balanced fulcrum. If you moved the anvil to a point 4 feet away from the pivot point, then $w \times d$ would be $300 \times 4 = 1{,}200$. This is greater than $W \times D$, which you calculated as 900. In this case since $wd > WD$, the plank will tilt to the left:

factoring

Sometimes expressions are easier to deal with when you "pull out" common factors from the terms. For example, if you had $6xy^2 + 3xy$, you could notice that each term is divisible by 3. Each term is also divisible by x. And each term is divisible by y as well.

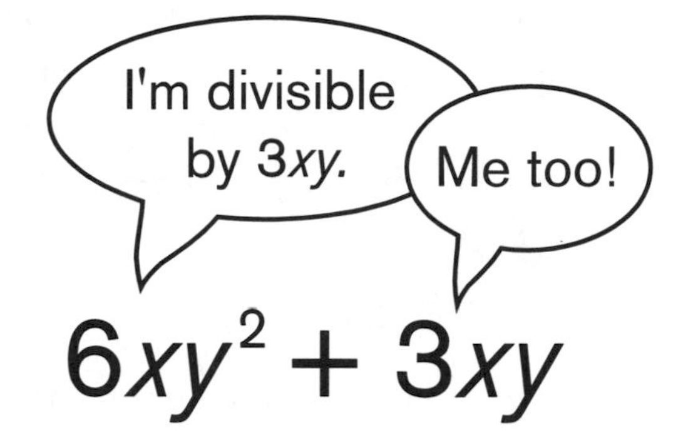

You can "pull out" a $3xy$ from each term to yield $3xy(2y + 1)$. You can check your work by distributing the $3xy$:

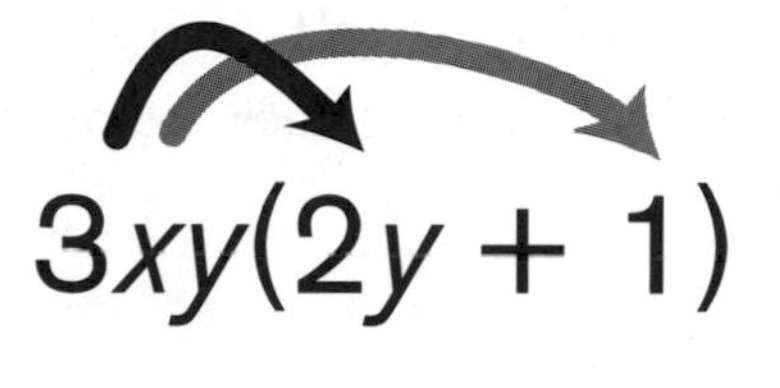

$$= 3xy \bullet 2y + 3xy \bullet 1$$

$$= 6xy^2 + 3xy$$

Exercise 9: Factor the expression $\mathbf{20x^2y + 15x + 10xy}$.

foil

"Foil" is an acronym for "First, Outer, Inner, Last, " which is the way in which you combine terms within two sets of parentheses. Let's look at the following question:

Which answer choices mathematically express the product of 2 more than x and 3 less than twice x?

a. $3x^2 + 7x + 6$
b. $3x^2 - 7x - 6$
c. $3x^2 + x - 6$
d. $3x^2 + x + 6$

You are asked to find the *product* so you know that you will be multiplying. What exactly are you multiplying? Well, one of the quantities given is "2 more than x," which is just $(x + 2)$. The second quantity given is "3 less than twice x," which can be expressed mathematically as $(2x - 3)$. When you multiply $(x + 2)$ by $(2x - 3)$, you get:

$$(x + 2)(2x - 3)$$

This would be a perfectly good answer except for one problem: It is not one of your choices! So after muttering comments about the question designer under your breath, you'll realize that you need to change the way you write your answer. Specifically, you must *expand* your current expression. You expand out $(x + 2)(2x - 3)$ by using FOIL. FOIL is just an acronym for **FIRST**, **OUTER**, **INNER**, and **LAST**. It describes the order in which you multiply your two sets of parentheses:

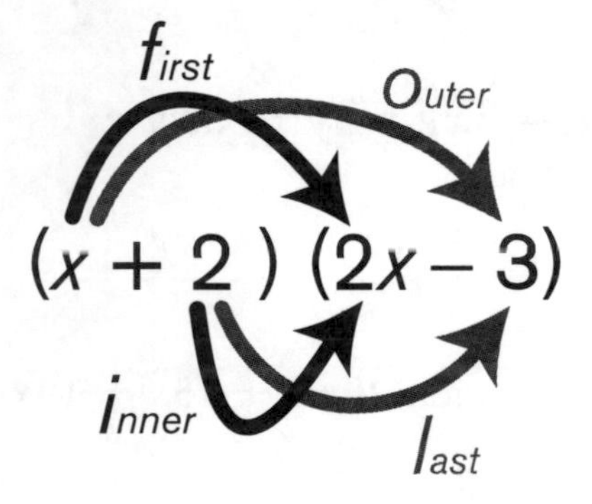

$(x + 2)(2x - 3) = 2x^2 - 3x + 4x - 6$. This simplifies to $3x^2 + x - 6$, which is choice **c**.

Exercise 10: Which answer mathematically expresses the product of five more than x and 1 more than twice x?

a. $x^2 + 11x + 5$
b. $2x^2 + x + 5$
c. $2x^2 + 10x + 5$
d. $2x^2 + 11x + 5$

reverse foil

In the above examples you used FOIL to expand expressions. Sometimes you will see an expression in expanded form, and you will need to perform a "reverse FOIL" in order to solve for your unknown. In other words, you are finding the factors of the given expression.

What are the solutions to $x^2 + 6x - 16 = 0$

The expression $x^2 + 6x - 16 = 0$ can be factored into two sets of parentheses:

$$(x \pm ?)(x \pm ?) = 0$$

Because the coefficient of the x^2 term is 1, you know that the sum of the 2 missing numbers is 6 (the coefficient of the x term) and the product of the 2 missing numbers is -16 (the lone number). The two numbers that satisfy these conditions are -2 and 8. Fill in the parentheses:

$$(x - 2)(x + 8) = 0$$

You have two quantities that, when multiplied, yield zero as the answer. Simply put, you have:

something × something = 0

If the answer is zero, then you know that one of those quantities (one of those "*somethings*") has to be zero. So you set both of those "*somethings*" equal to 0.

$$(x - 2)(x + 8) = 0$$

$$\begin{array}{r|l} x - 2 = 0 & x + 8 = 0 \\ x = 2 & x = -8 \end{array}$$

Thus, the solutions are $x = 2$ and $x = -8$.

Exercise 11: If x is a positive number, and $x^2 - 8x - 9 = 0$, what is the value of x?

a. -1
b. 3
c. 9
d. both b & c
e. both a & c

simultaneous equations

When you are dealing with equations that have more than one variable, instead of isolating your variable, sometimes it is easier to use simultaneous equations. All you need to do is arrange your two equations on top of one another. Be sure that the like terms are written in the same order. You can add the equations together, or you can subtract them. If you are strategic, you can figure out what to do in order to calculate the answer to the question. Always pay close attention to what the question "wants."

Let's look at an example:

If $2x + y = 13$, and $5x - y = 1$, what is the value of x?
a. -2
b. 1
c. 2
d. 3

Let's set up a *simultaneous equation*. Keep in mind that the question asks us for the value of x. Arrange your two equations on top of each other, and in this case, you will add them together:

$$\begin{array}{r} 2x + y = 13 \\ + \; 5x - y = \;\; 1 \\ \hline 7x = 14 \\ x = \;\; 2 \end{array}$$

Note that adding them together was a good choice, because you got rid of the y variables.

$$\begin{array}{r} 2x + y = 13 \\ + \; 5x - y = \;\;1 \\ \hline 7x = 14 \\ x = \;\;2 \end{array}$$

Now you know $x = 2$, choice **c**.

Exercise 12: If $3x + y = 40$ and $x - 2y = 4$, what is the value of y?

a. $-5\frac{3}{5}$
b. 4
c. $5\frac{3}{5}$
d. 8

Hint: You want to find the value of y, so try to get rid of x.

solutions to chapter exercises

Exercise 1: Combine like terms:

$$2x + 3x^2 + 5x - x^2 + x + 7x^3 =$$

$$\mathbf{2x} + 3x^2 + \mathbf{5x} - x^2 + \mathbf{x} + 7x^3 =$$

$$\mathbf{8x} + 3x^2 - x^2 + 7x^3 =$$

$$8x + \mathbf{3x^2 - x^2} + 7x^3 =$$

$$8x + 2x^2 + 7x^3$$

Exercise 2: d. Presented with the question:

$(\frac{x}{3}) + (\frac{3x}{10}) - (\frac{2x}{5})$ is equivalent to

a. $\frac{7x}{15}$
b. $\frac{31x}{30}$
c. $\frac{8x}{18}$
d. $\frac{7x}{30}$

First, notice that all the answer choices are expressed as fractions. That's your clue that you need to find a common denominator for these terms. The denominators are 3, 10, and 5, so 30 would be a great common denominator. To turn $\frac{x}{3}$ into something over 30, you are multiplying top and bottom by 10. This first term becomes $\frac{10x}{30}$. Now look at the second term, $\frac{3x}{10}$. To turn this term into something over 30, you'll multiply top and bottom by 3, yielding $\frac{9x}{30}$. Next, to turn $\frac{2x}{5}$ into something over 30, multiply top and bottom by 6, and you get $\frac{12x}{30}$. The original expression $(\frac{x}{3}) + (\frac{3x}{10}) - (\frac{2x}{5})$ is now $(\frac{10x}{30}) + (\frac{9x}{30}) - (\frac{12x}{30}) = \frac{7x}{30}$.

Exercise 3: Given $7x + 2 = 5x + 14$, you are told to find the value of x.

The first thing you want to do is isolate your variable. This means you want to combine your x terms on one side of the equation and your numbers on the other side of the equation. Below you will subtract $5x$ from both sides in order to combine x terms:

$$\begin{array}{rr} 7x + 2 = & 5x + 14 \\ -\,5x \quad & -\,5x \quad \\ \hline 2x + 2 = & 14 \end{array}$$

Now you will subtract 2 from both sides in order to isolate the x term.

$$\begin{array}{rr} 2x + 2 = & 14 \\ -\,2 \quad & -\,2 \\ \hline 2x \quad = & 12 \end{array}$$

Finally, divide both sides by 2 to get $x = 6$.

Exercise 4: First, simplify the expression $x^2 + 12 \leq 4^2$ by squaring the 4:

$$x^2 + 12 \leq 16$$

$$\begin{array}{rr} x^2 + 12 \leq & 16 \\ -\,12 \quad & -\,12 \\ \hline x^2 \quad \leq & 4 \end{array}$$ Subtract 12 from both sides.

$-2 \leq x \leq 2$ Numbers between -2 and 2 (inclusive) will yield numbers less than or equal to 4 when they are squared.

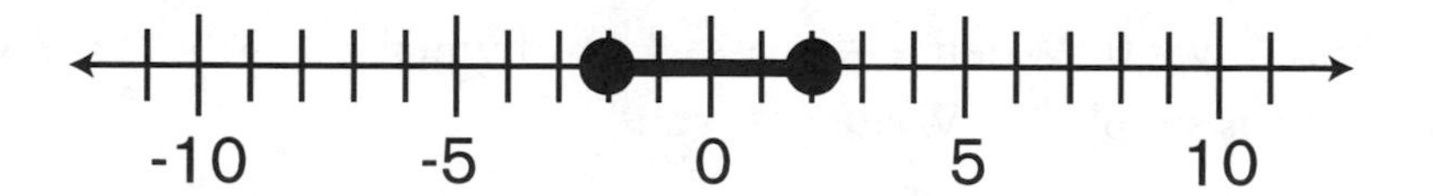

Exercise 5: Let's look at the phrase:

Erik has 3 CDs less than Danny.

3 CDs less than Danny would be $D - 3$. If Erik had 3 CDs less than Danny, you know:

Danny's – 3 = Erik's

So, Erik has $E = D - 3$, choice **d**.

Exercise 6: To calculate the value of the expression $5x^2 + 2xy^3$ when $x = 3$ and $y = -2$, you just substitute the given values for the variables. Substituting $x = 3$ and $y = -2$ into the equation $5x^2 + 2xy^3$ you get: $5(3)^2 + 2(3)(-2)^3 = (5)(9) + (2)(3)(-8) = 45 + (6)(-8) = 45 + (-48) = -3$, or choice **a**.

Exercise 7:

x	y
0	0
1	–
2	4
3	–
4	16

When x is 0, y is 0. When $x = 2$, $y = 4$. And when $x = 4$, $y = 16$. What is the pattern? y is simply x^2. Write $y = x^2$.

Exercise 8: You know Train 1 is traveling eastbound at 60 mph and Train 2 is traveling at 50 mph in an westbound direction. The trains were initially

220 miles apart. We need to figure out how long it will be before they pass each other.

First, draw a diagram:

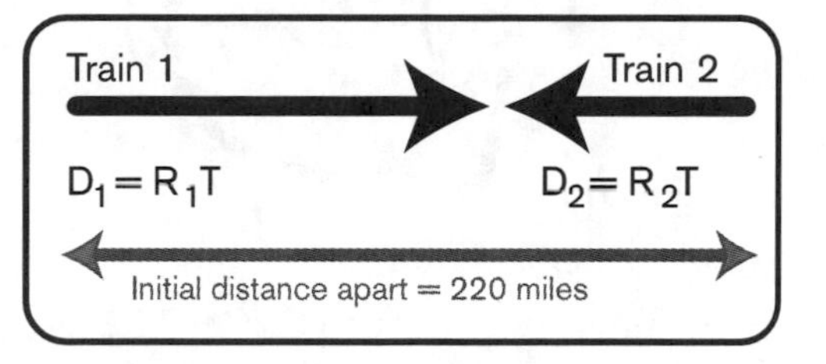

The initial distance between the 2 stations will equal the sum of the distances traveled by each train for the period of time in question.

$$\text{Total Distance} = \mathbf{D_1 + D_2}$$

$$220 \text{ miles} = \mathbf{R_1T + R_2T}$$

$$220 = (60)(t) + (50)(t)$$

$$220 = 110t$$

$$2 = t$$

It will take 2 hours.

Exercise 9: Each term in the expression $\mathbf{20x^2y + 15x + 10xy}$ is divisible by $5x$. You pull out a $5x$ to yield:

$$\mathbf{5x(4xy + 3x + 2y)}$$

Exercise 10: d. The fact that you see x^2 in every answer choice is a clue that FOIL needs to be used. So what goes in each set of parentheses? The first pair of parentheses will be filled by "five more than x," which is represented mathematically as $(x + 5)$. The second pair of parentheses will be filled by "1 more than twice x," which is simply $(2x + 1)$. The question asks you to calculate a product, thus you multiply using FOIL:

$$\mathbf{(x + 5)(2x + 1) =}$$

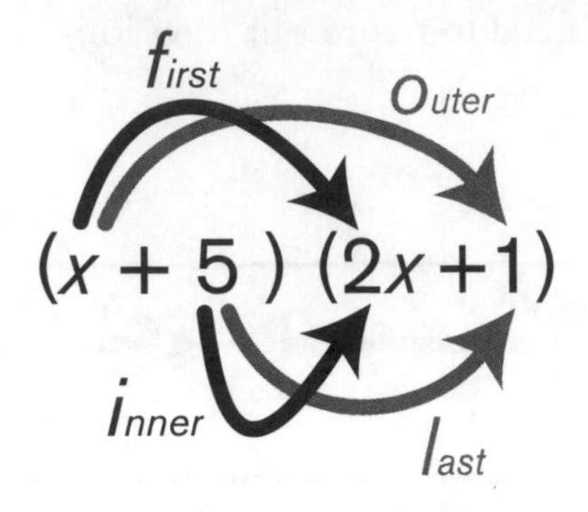

$= 2x^2 + x + 10x + 5$. This simplifies to $2x^2 + 11x + 5$.

Exercise 11: Given x is a positive number, and $x^2 - 8x - 9 = 0$, you can actually cross off choices **a** and **e** because you know that x is positive.

a. -1
b. 3
c. 9
d. both b & c
e. both a & c

Next, the expression $x^2 - 8x - 9 = 0$ can be factored into two sets of parentheses:

$$(x \pm ?)(x \pm ?) = 0$$

Again, because the coefficient of the x^2 term is 1, you know that the sum of the 2 missing numbers is -8 (the coefficient of the x term) and the product of the 2 missing numbers is -9 (the lone number).

The two numbers that satisfy these conditions are -9 and 1. Fill in the parentheses and set each set of parentheses equal to 0:

$$(x - 9)(x + 1) = 0$$

$x - 9 = 0$	$x + 1 = 0$
$x = 9$	$x = -1$

But be careful! The question told us that x is positive! This means that only 9 is correct, choice **c**.

Exercise 12: b. Here you will set up a simultaneous equation. Note that the question asks you to find y, so it would be nice to make a simultaneous equation that allows you to cross out x. The first given equation has a $3x$ in it. The second equation has an x in it. If you multiply the entire second equation by 3, you will be able to subtract it from the first and there will be no more x term! First, multiply $x - 2y = 4$ by 3 to get $3x - 6y = 12$. Next, let's line up our equations:

$$\begin{array}{r} 3x + y = 40 \\ -(3x - 6y = 12) \\ \hline 7y = 28 \\ y = 4 \end{array}$$

Note that you got $7y$ because y minus a negative $6y$ = y plus $6y$.

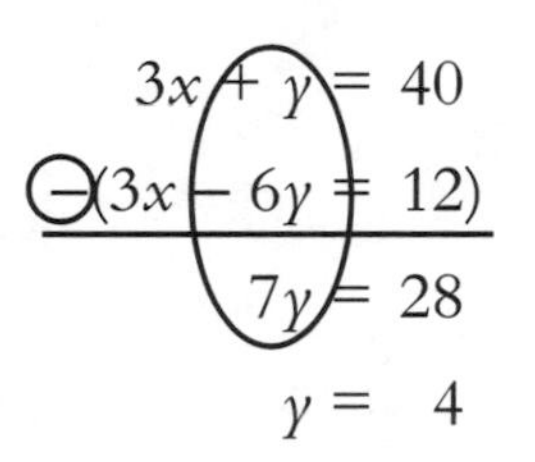

$y = 4$

Geometry and Measurement

the metric system

The metric system is a very well thought out way to measure distances, volumes, and masses. Once you understand what the prefixes mean and what the basic terms used for measurement are, you will have no problem dealing with these units. For example, meters (m) are used to measure length.

1 m = 1,000 millimeters = 1,000 mm

1 m = 100 centimeters = 100 cm

1 m = 10 decimeters = 10 dm

Here are other prefixes you will see:

The prefix . . .	Means . . .	Example . . .
milli	$\frac{1}{1,000}$ of	1 milliliter is $\frac{1}{1,000}$ of a liter
centi	$\frac{1}{100}$ of	1 centimeter is $\frac{1}{100}$ of a meter
deci	$\frac{1}{10}$ of	1 decigram is $\frac{1}{10}$ of a gram
deca	10 times	1 decameter is 10 meters
hecto	100 times	1 hectoliter is 100 liters
kilo	1,000 times	1 kilometer is 1,000 meters

So what is a liter, a meter, and a gram? A liter is used to measure volume. A gram is used to measure mass. And a meter is used to measure length.

Term	Used to measure
Liter	Volume
Gram	Mass
Meter	Length

advanced conversions

1 acre = 43,560 square feet
1 liter = 1,000 cubic centimeters

You may already be used to the customary system. Here are some units that you should know:

Customary Units	
1 foot = 12 inches	1 cup = 8 fluid ounces
3 feet = 1 yard	1 pint = 2 cups
1 mile = 5,280 feet	1 quart = 2 pints
1 pound = 16 ounces	1 gallon = 4 quarts
1 ton = 2,000 pounds	

operations with mixed measures

A mixed measure is part one unit and part another unit. For example, "12 feet 5 inches" is part *feet* and part *inches*. When adding or subtracting mixed measures, you just need to align the units and perform the operation at hand. You can then **rename** units as necessary.

Example: Ryan's band played for 1 hour and 35 minutes. Ray's band played for 1 hour and 40 minutes. When combined, the two performances lasted how long?

$$\begin{array}{r} 1 \text{ hr} + 35 \text{ min} \\ + \; 1 \text{ hr} + 40 \text{ min} \\ \hline 2 \text{ hr} + 75 \text{ min} \end{array}$$

Rewrite 75 minutes as 60 min + 15 min.

2 hr + (60 min + 15 min)

Rename 60 minutes as 1 hour.

2 hr + 1 hr + 15 min

3 hr + 15 min

using proportions to convert units

Proportions can be used to convert units. Look at the equivalents below:

1 cup = 8 ounces	1 hour = 60 minutes
1 meter = 100 cm	1 min = 60 seconds

If you know that 1 cup equals 8 ounces, you can easily figure out how many cups are in 64 ounces by setting up a proportion:

$$\frac{1 \text{ cup}}{8 \text{ oz}} = \frac{? \text{ cup}}{64 \text{ oz}}$$

Cross multiplying, you get 64 • 1 = 8 • ?, or 64 = 8 • ?. Dividing both sides by 8 you get ? = 8.

Exercise 1: Chris lays two planks of wood end to end. If one plank is 6 yd., 2 ft., 8 in. long, and the other is 7 yd., 1 ft., 5 in. long, how long are they when combined?

rays and angles

Terminology:

- A **ray** is part of a line that has one endpoint. (It extends indefinitely in one direction.)
- An **acute angle** is less than 90°.
- A **right angle** equals 90°.
- A **straight line** is 180°.
- An **obtuse angle** is greater than 90°.
- Two angles are **supplementary** if they add to 180°.
- Two angles are **complementary** if they add to 90°.
- If you **bisect** an angle, you cut it exactly in half.
- **Vertical angles** are formed when two lines intersect; the opposite angles are equal.

Exercise 2: Refer to the figure below when completing the following chart.

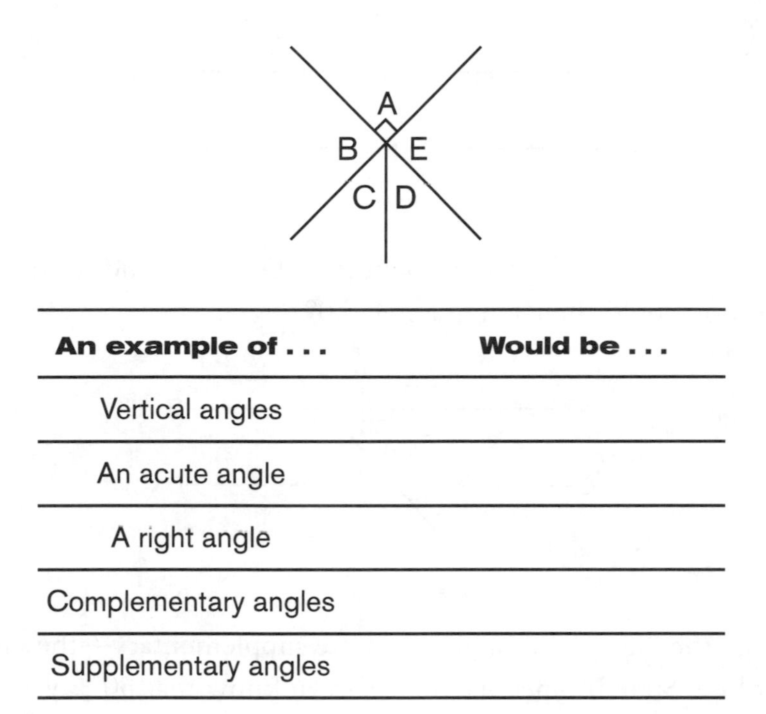

An example of . . .	Would be . . .
Vertical angles	
An acute angle	
A right angle	
Complementary angles	
Supplementary angles	

parallel lines

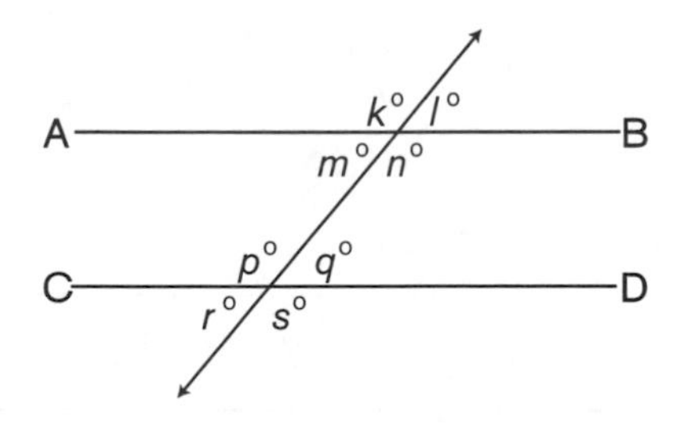

Facts:

- A line that crosses the pair of parallel lines will generate **corresponding angles** that are equal. (For example, angles p and k are in corresponding positions, so they are equal.)
- **Alternate interior angles** are equal. (m & q are alternate interior angles, so they are equal. This is also true for p and n.)
- **Alternate exterior angles** are equal. (This means $k = s$, and $r = l$.)

Example: If line segments $\overline{AB}$ and $\overline{CD}$ are parallel, what is the value of y?

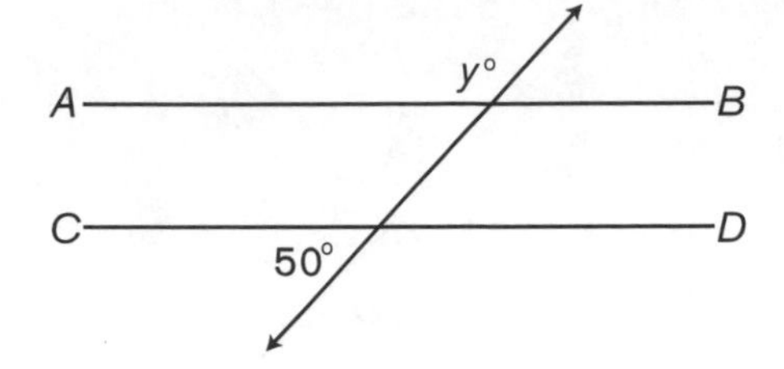

Here, you see that the 50° angle beneath $\overline{CD}$ corresponds to the angle beneath $\overline{AB}$. You could then add in a new 50° angle:

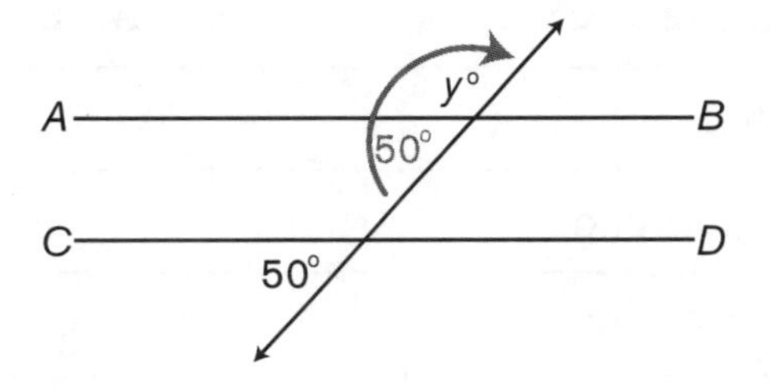

You see that the "new" 50° angle and y are supplementary—they form a straight line. Straight lines are 180°, so you know that $50 + y = 180$. Subtracting 50 from both sides yields $y=130°$.

Exercise 3: If, in the figure below $\overline{OP}$, $\overline{QR}$, $\overline{ST}$ and $\overline{UV}$ are parallel, what is the value of angle b?

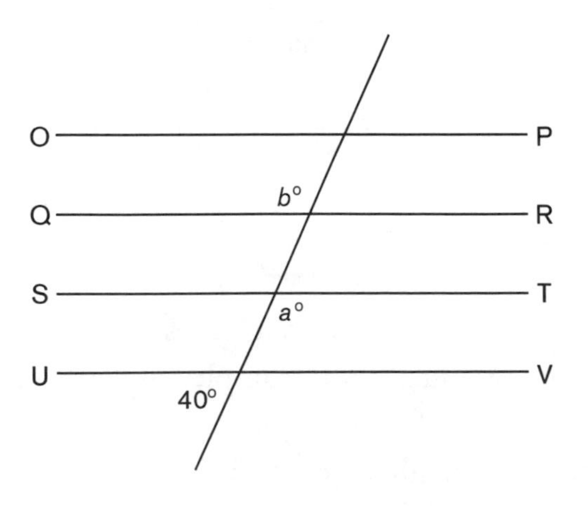

interior angles of common shapes

The interior angles of a **triangle** add up to 180°.

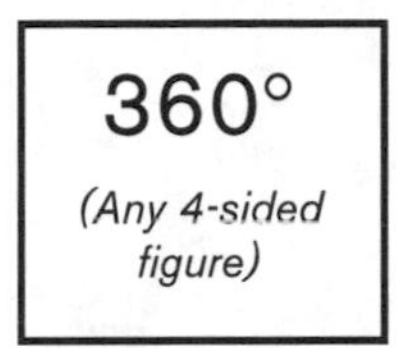

The interior angles of **any** four-sided figure, or **quadrilateral**, add up to 360°.

360°

(Any 4-sided figure)

- square
- rectangle
- parallelogram
- rhombus
- trapezoid

The central angles of a **circle** add up to 360°.

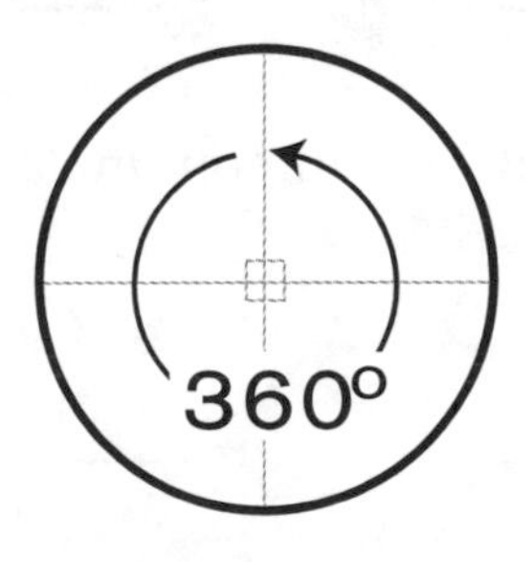

For a **polygon with n sides**, the interior angles will add to **180°(n − 2)**.

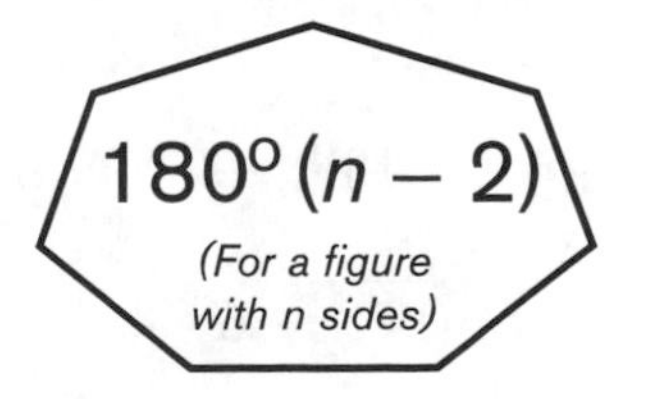

Exercise 4: Calculate the value for the sum of the interior angles in an octagon.

congruence

When two figures are *congruent*, they are exactly the same.

similarity

Similar figures must have:

- matching angles that are congruent
- matching sides that are proportional

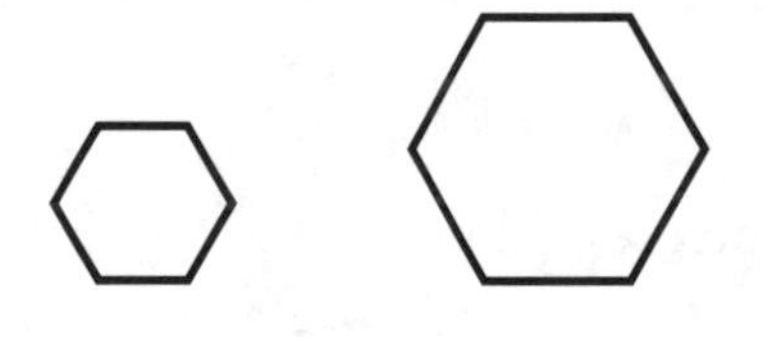

Similar triangles are the most common similar figures that you will see:

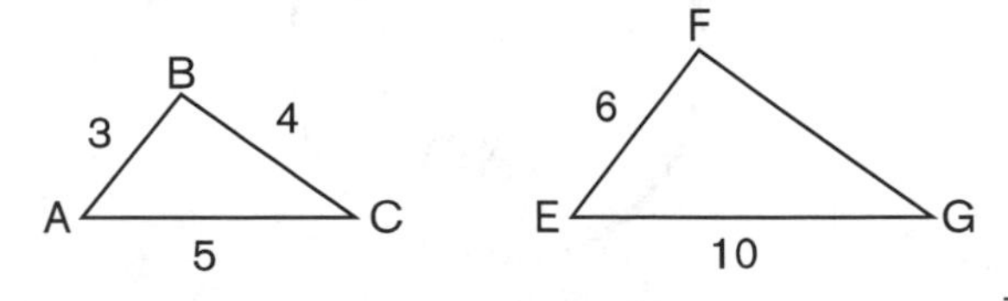

You can set up a proportion to solve for the missing side, $\overline{FG}$.

$$\frac{3}{6} = \frac{4}{\overline{FG}}$$

Cross-multiplying, you get:

$$3(\overline{FG}) = 6 \cdot 4$$

$$3(\overline{FG}) = 24$$

Next, you divide both sides by 3 to get:

$$\overline{FG} = 8$$

Exercise 5: Given that $\overline{BC}$ is parallel to $\overline{DE}$, if the ratio of $\overline{AB}:\overline{BD}$ is 2:3, then calculate the ratio of $\overline{BC}:\overline{DE}$.

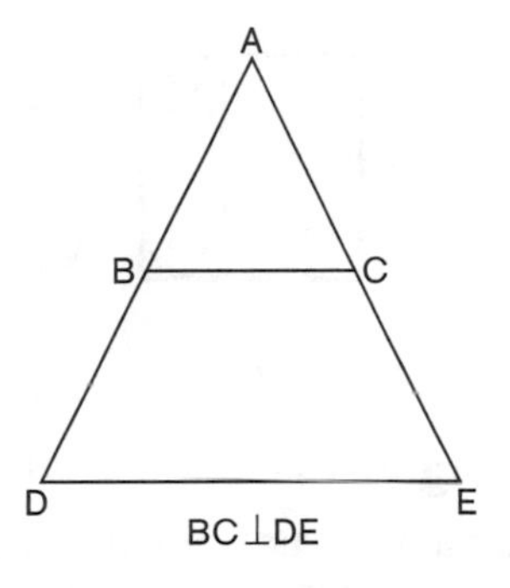

symmetry

Shapes are said to be symmetrical if you can draw a line through that shape, forming 2 halves that are mirror images of each other.

This hexagon has a line of symmetry:

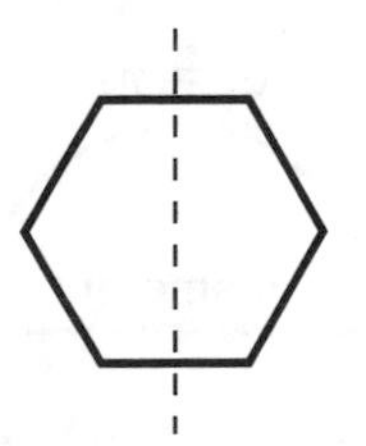

This quadrilateral does not have a line of symmetry. Notice that no matter how you draw the dashed line, you will not make two mirror-image halves:

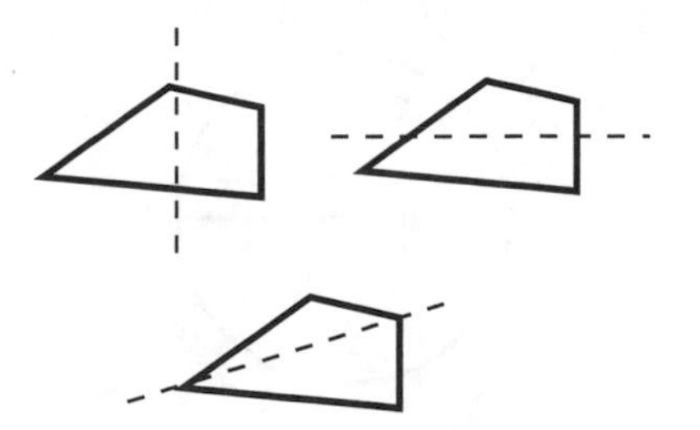

perimeter and circumference

When we measure the distance around a noncircular shape, we call it the **perimeter**.

When we measure the distance around a circle, it is called a **circumference**.

The perimeter of a **square** is equal to $4s$, where s is the length of a side. Since all four sides are equal, when you measure the distance around a square, you get $s + s + s + s$, or $4s$.

$4s$

The perimeter of a **rectangle** is 2L + 2W, where L equals the length and the W equals the width. Always remember that the length is longer.

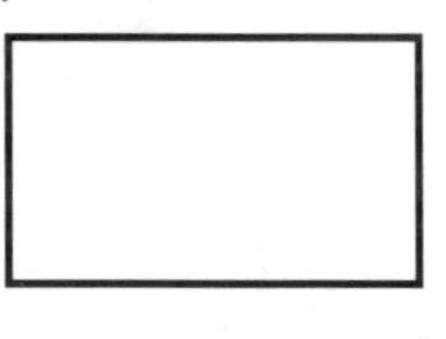

2L + 2W

The circumference of a **circle**, C, equals πd, or $2\pi r$.

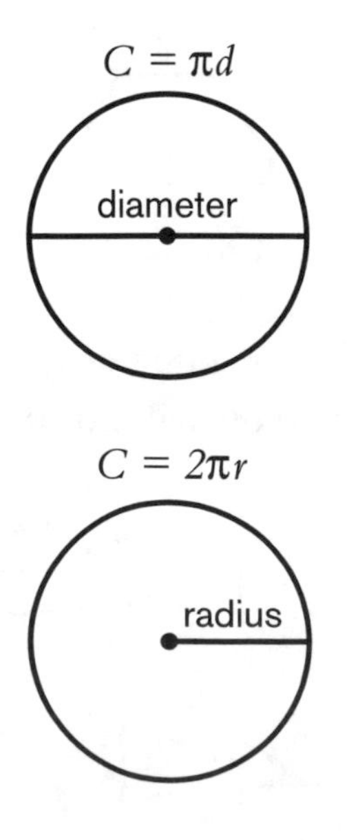

Notice that both of these formulas make sense because $2r = d$.

To find the perimeter of a triangle, you just add up the lengths of all three sides:

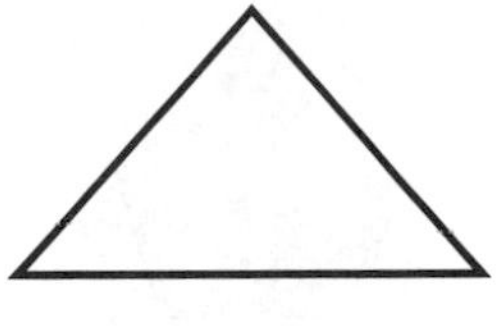

Add up sides.

To find the perimeter of a regular polygon, you just need to know one side. This is because regular polygons have all sides equal. For example, a regular hexagon has six equal sides:

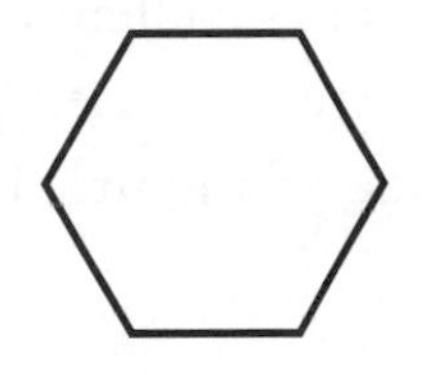

Add up sides.

If a figure is not regular, sometimes markings will denote congruent sides:

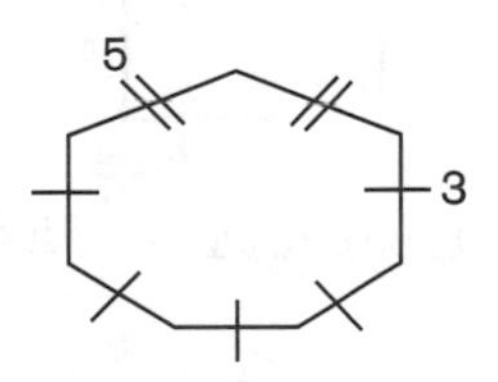

The sides with the same markings are equal:

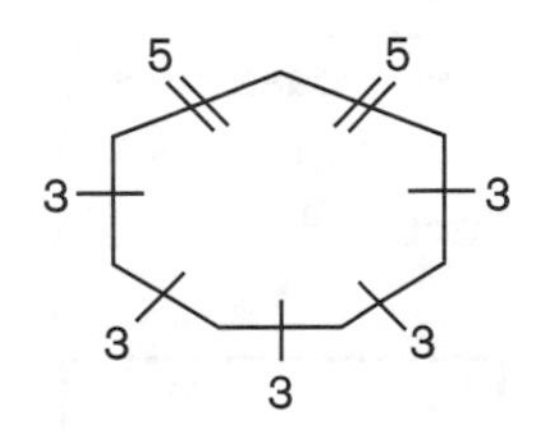

The perimeter is then $3 \cdot 5 + 5 \cdot 2 = 15 + 10 = 25$ units.

Exercise 6: If the figure below is a regular pentagon with a side equal to 5mm, calculate its perimeter.

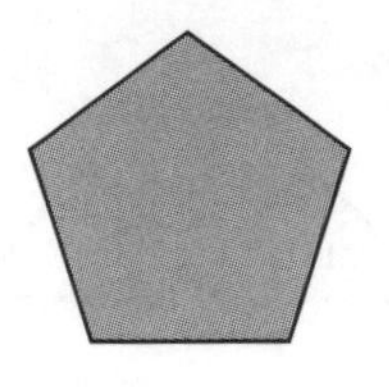

area

When you measured lengths, you used different units, like feet, meters, inches, and so on. When you find the area, you use square units. The area measures the amount of square units inside of a figure. Here are the area formulas that you should know:

Square $A = s^2$

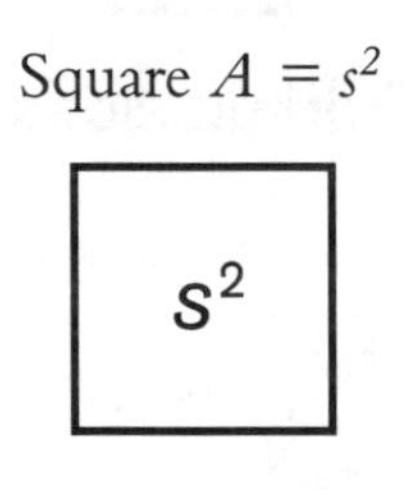

Triangle $A = \frac{1}{2}bh$

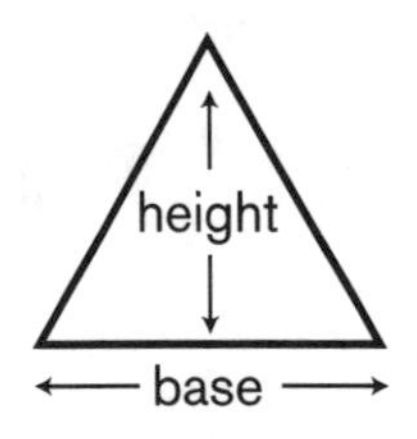

Rectangle A = LW

L x W

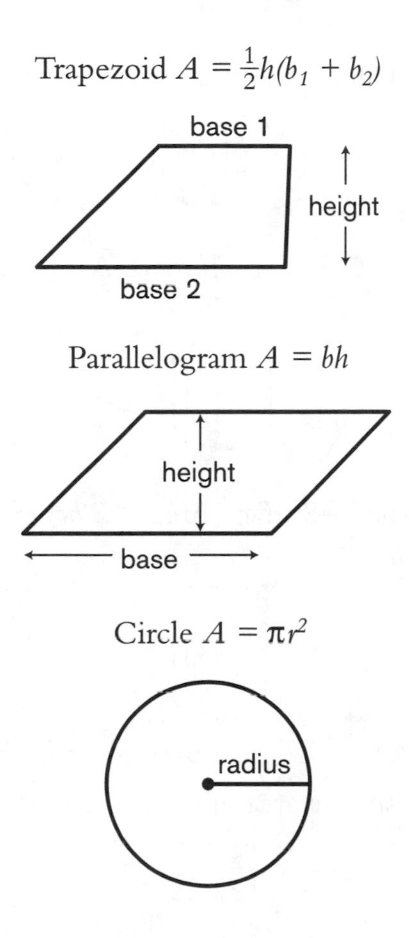

Exercise 7: If the length of the rectangle below is doubled, how much greater will its area be?

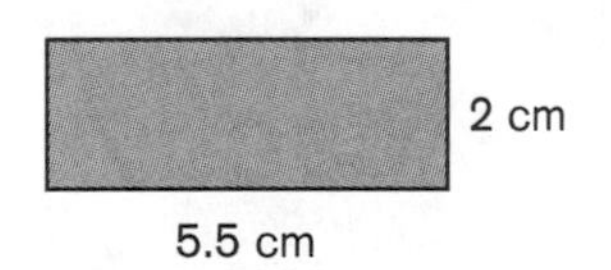

surface area

The surface area is just the area on the surface of a solid figure. Surface area is also measured in units^2 (square units). In order to find the surface area of a cube, you would find the area of each face and then find the total of all of the faces. Since the area of a square is s^2, you will have six faces with an area of s^2, so in all you will have $6s^2$.

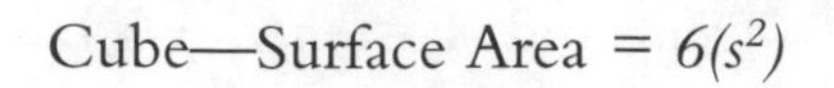

Cube—Surface Area = $6(s^2)$

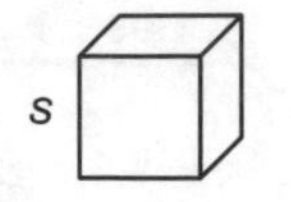

Right Circular Cylinder—Surface Area = $2\pi rh + 2\pi r^2$

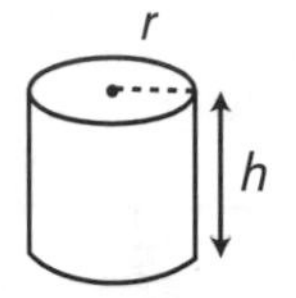

Rectangular Solid—Surface Area = $2(lw) + 2(hw) + 2(lh)$

Exercise 8: What is the surface area of the rectangular solid below?

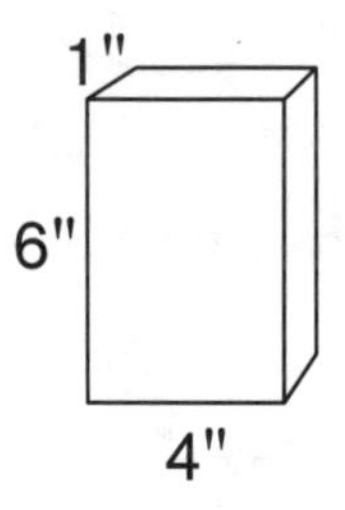

volume

The volume measures the amount of cubic units needed to fill a three-dimensional solid. As you might have guessed, volumes are measured in units^3 (cubic units).

Cube	$V = s^3$
Right Circular Cylinder	$V = \pi r^2 h$
Rectangular Solid	$V = lwh$

Exercise 9: If the barrel below is filled to a height of four feet with oil, what is the volume this oil occupies inside the barrel?

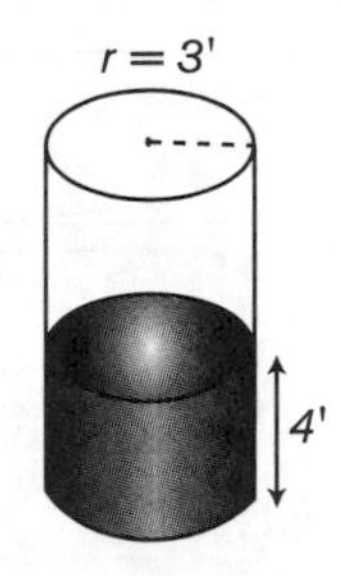

a closer look at triangles

Triangle Terminology:

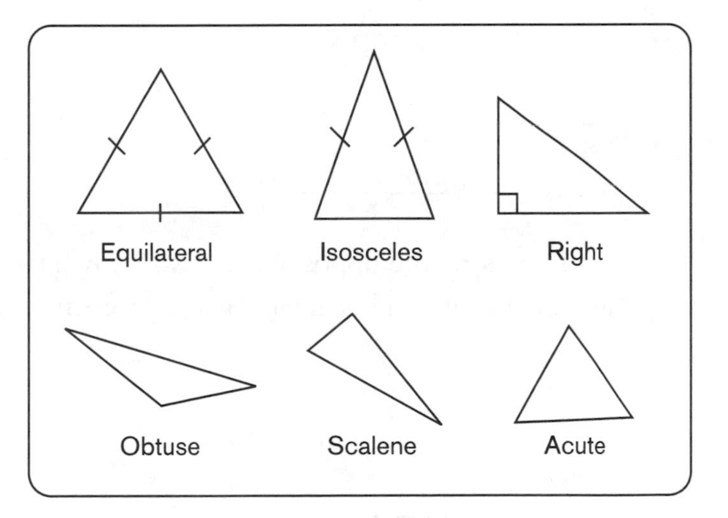

Equilateral triangles have all angles equal to 60°. All sides of equilateral triangles are congruent (equal).

Obtuse triangles have one angle that is greater than 90°.

Isosceles triangles have two congruent sides (and the angles opposite these equal sides are equal as well).

Scalene triangles have no sides that are congruent.

Right triangles have one right (90°) angle.

Acute triangles have all angles less than 90°.

Triangle Facts:

1. **Triangle Inequality Theorem:** A side of a triangle is always less than the sum of the other two sides (and greater than their difference).

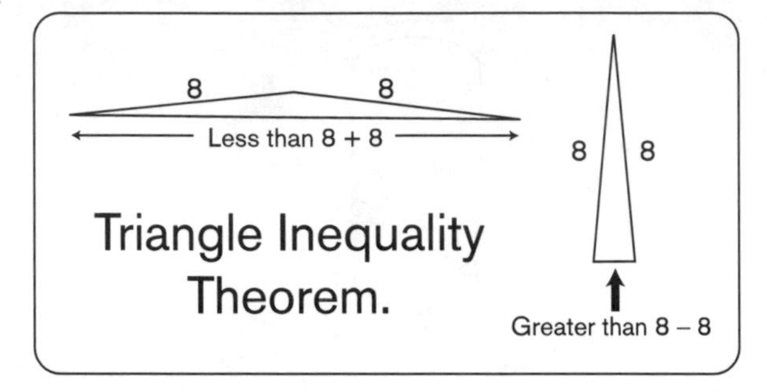

2. Given a triangle, congruent sides are opposite congruent angles. In other words, equal sides are opposite equal angles.

   Example: To find the missing side below, you can first calculate the opposite angle:

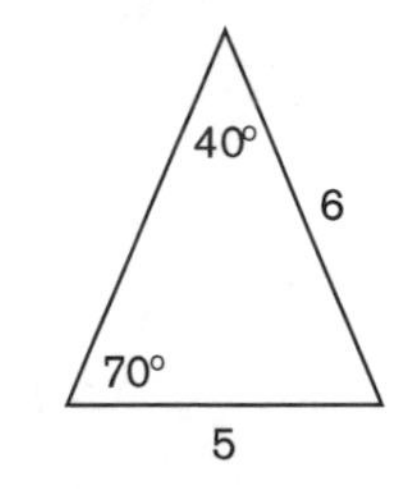

   40 + 70 = 110. Therefore, the opposite angle is 70°. In a triangle, equal sides are opposite equal angles. This means that the missing side is also 6.

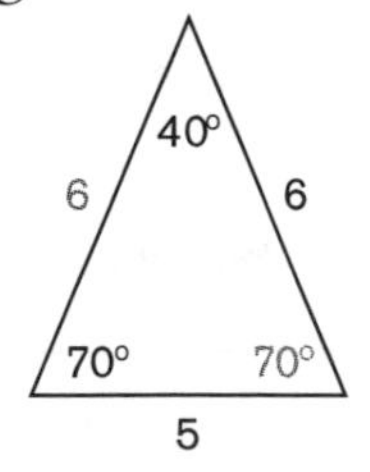

3. Given a triangle, larger sides are opposite larger angles, and smaller sides are opposite smaller angles.

   Example: In the triangle below, which side is the largest?

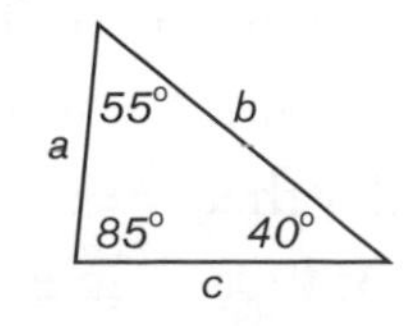

   Side *b* is opposite the largest angle, so it must be the largest side.

Right Triangles:
Right triangles are special because you can use the Pythagorean Theorem. The Pythagorean Theorem is:

$$a^2 + b^2 = c^2$$

where a and b are the legs,
and c is the hypotenuse

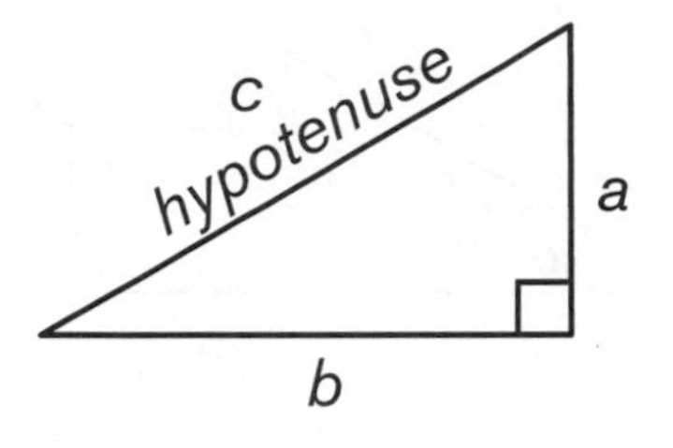

Exercise 10: A flat wooden board is leaning against a building as shown below. Calculate the horizontal distance from the bottom of the board to the base of the building.

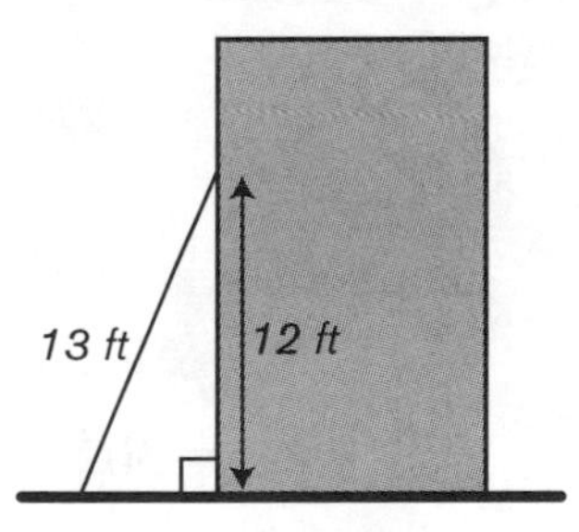

triangle trigonometry

There are special ratios you can use with right triangles. They are based on the trigonometric functions called *sine, cosine,* and *tangent.* The popular mnemonic to use is:

SOH CAH TOA

For an angle, θ, within a right triangle, you can use these formulas:

$$\sin \theta = \frac{\text{opposite side}}{\text{hypotenuse}}$$

$$\cos \theta = \frac{\text{adjacent side}}{\text{hypotenuse}}$$

$$\tan \theta = \frac{\text{opposite side}}{\text{adjacent side}}$$

Let's look at the figure below:

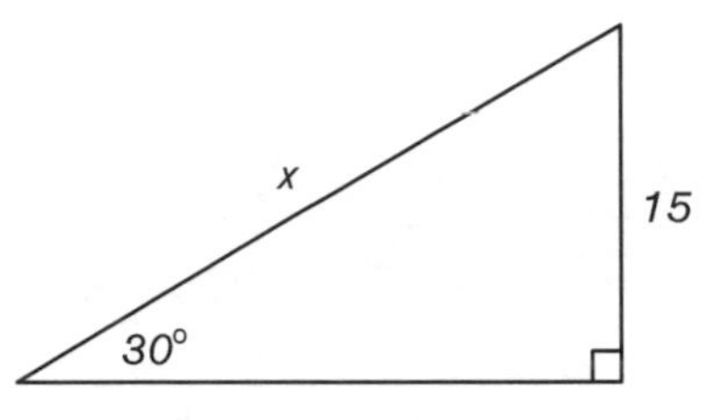

Notice that the leg of the right triangle marked as 15 is **opposite** the 30° angle. The hypotenuse is marked as x. In order to find x, you can use the ratio $\sin = \frac{\text{opp}}{\text{hypot}}$.

By using a calculator, our memory, or a reference chart, you can calculate that $\sin 30 = \frac{1}{2}$. Thus, you can solve for x.

$$\sin 30 = \frac{\text{opposite side}}{\text{hypotenuse}} = \frac{15}{x}$$

$$\frac{1}{2} = \frac{15}{x}$$

$$x = 15 \bullet 2 = 30$$

a closer look at quadrilaterals

Quadrilateral Terminology and Facts:

- **Trapezoids** have exactly *one* pair of parallel sides.
- **Parallelograms** have *two* pairs of parallel sides. The opposite sides are congruent. The opposite angles are congruent. The diagonals of parallelograms bisect each other.

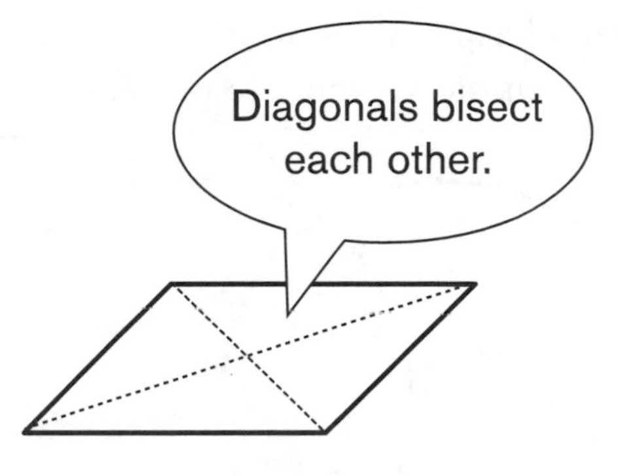

- **Rectangles** are parallelograms with *four right angles*. This means that the diagonals of a rectangle bisect each other.
- A **rhombus** is a parallelogram with *four congruent sides*. The diagonals of a rhombus not only bisect each other, they bisect the angles that they connect as well! Also, the diagonals are perpendicular.

Diagonals bisect
each other & are
perpendicular!

- A **square** is a *rhombus with four right angles*.

Diagonals bisect
each other & are
perpendicular!

Exercise 11: In the Venn Diagram below, fill in the terms: Rhombus, Rectangle, and Square in the appropriate positions.

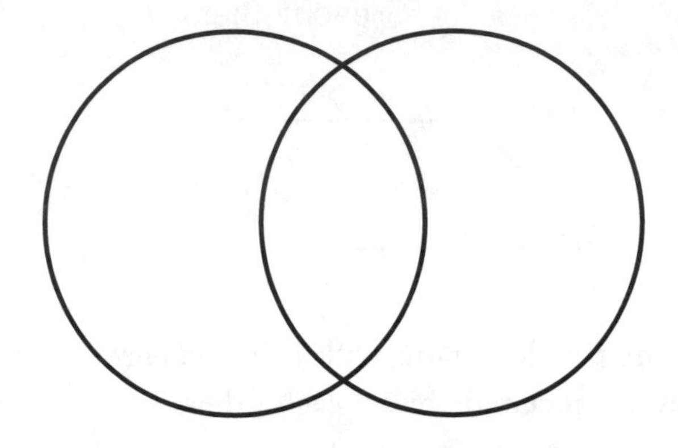

a closer look at circles

Circle Terminology:

- A **chord** is a line segment which joins two points on a circle.
- A **radius** is a line segment from the center of the circle to a point on the circle.
- A **diameter** is a chord which passes through the center of the circle and has end points on the circle.
- A **minor arc** is the smaller curve between two points.
- A **major arc** is the larger curve between two points.

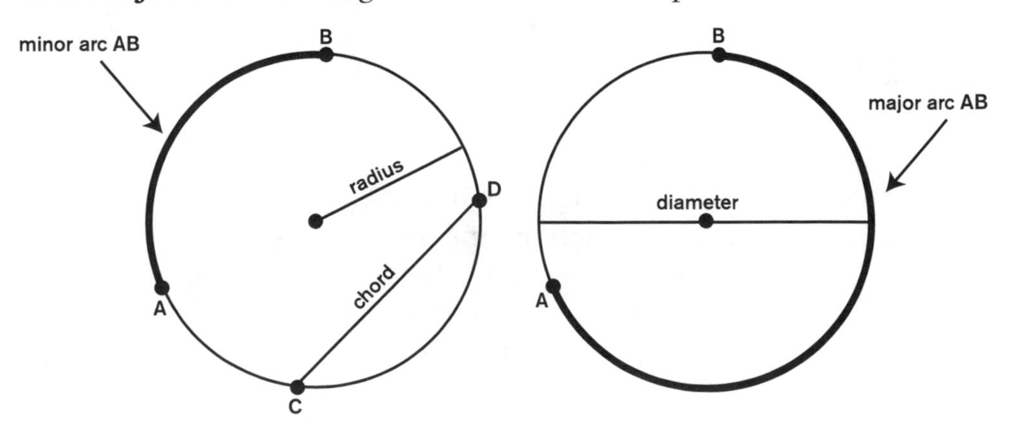

Tips on dealing with circles:

- Questions that involve the **rolling** of a wheel are usually circumference questions in disguise. As the wheel rolls, the distance around (circumference of) the wheel is in contact with the ground. Questions that ask, "How many times will the wheel rotate?" are really based on circumference.

- You can use the angle measure of a shaded "slice" of a circle to determine its area. If the shaded slice has a measure of 90°, you know that this is $\frac{1}{4}$ of the circle. The area of the slice would then be $\frac{1}{4}$ of the area of the whole circle. A shaded "slice" that has an angle of 60° would be $\frac{1}{6}$ of the total area, and so forth.

Exercise 12: Which distance below represents the number of feet traveled by the wheel pictured below as it revolves 20 times along its path?

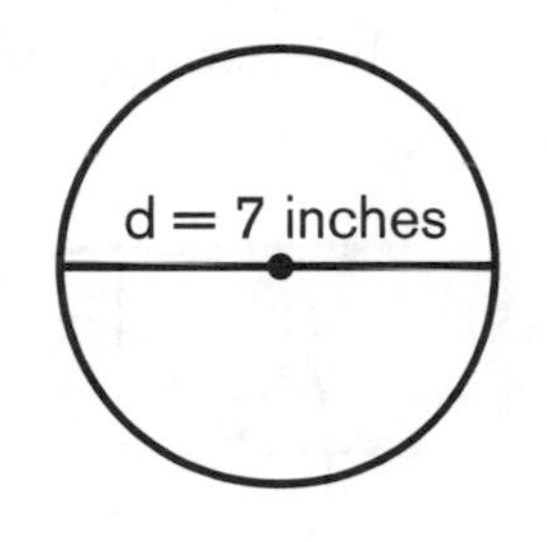

a. $36\frac{2}{3}$

b. 40

c. $42\frac{1}{3}$

d. 45

shaded areas

Sometimes you will need to calculate shaded areas of a given figure. These types of calculations require logical thinking, some analysis, and visualization as well.

For example, let's calculate the area of the shaded region below.

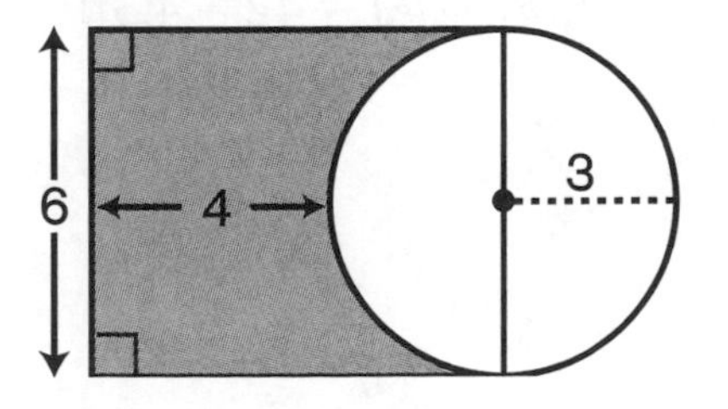

This question requires the knowledge of two area formulas:

- Area of rectangle = length × width
- Area of circle = πr^2

This question also requires some reasoning. Exactly how much of the whole figure is shaded? How can you use these area formulas to help? Well, you might've noticed that the shaded region is just the area of the rectangle minus the area of $\frac{1}{2}$ the circle. You can write a formula for yourself:

$$\textit{Area shaded} = \textit{Area of Rectangle} - \frac{1}{2}\textit{Area of Circle}$$

Let's get all the pieces you need by marking up the figure a different way:

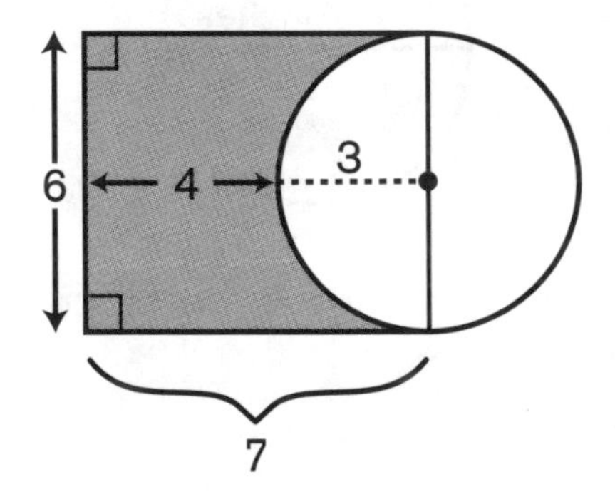

Notice that by drawing a new radius, you know that the length of the rectangle is 7. You already knew that the width was 6, so the area of the rectangle is just length × width = 7 × 6 = 42.

Now, you will figure out the area of the circle using $A = \pi r^2$, which becomes $\pi(3)^2 = 9\pi$. If the area of the whole circle is 9π, then the area of half the circle will be $\frac{1}{2} \times 9\pi = 4.5\pi$.

Thus, the area of the shaded region is

$$\textit{Area shaded} = \textit{Area of Rectangle} - \frac{1}{2}\textit{Area of Circle}$$
$$\textit{Area shaded} = 42 - 4.5\pi$$

Exercise 13: If the side of the square below is 16, what is the area of the shaded region?

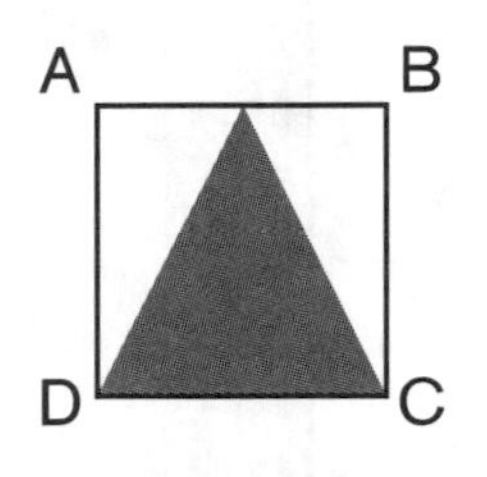

coordinate geometry

A coordinate grid is used to plot or locate points on a plane. The horizontal axis is called the *x*-axis, and the vertical axis is called the *y*-axis. There are four quadrants created by the intersection of the *x* and *y*-axis.

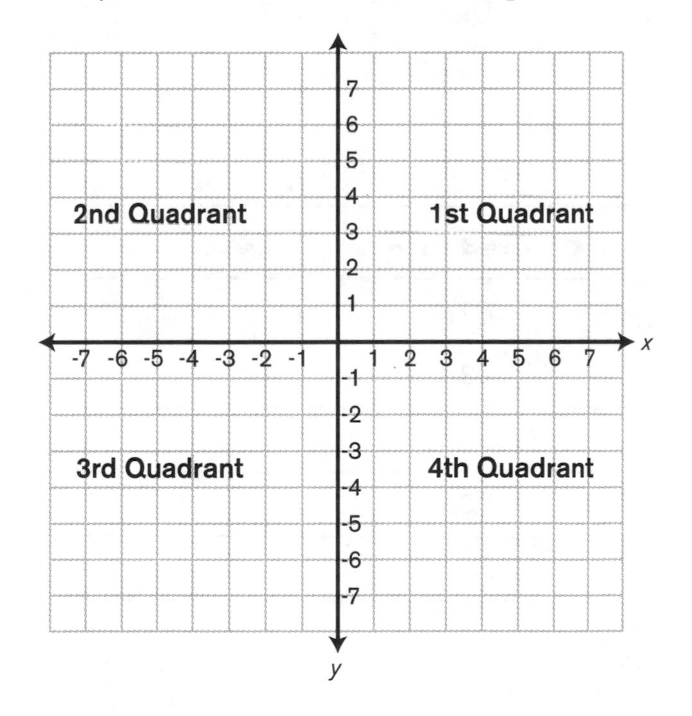

It is easy to plot points on the coordinate grid. Points are given as **ordered pairs,** which are just x/y pairs. An ordered pair is always written (x, y). You simply go left or right along the *x*-axis to find your *x*, and then you go up or down on to the appropriate *y*-coordinate.

Exercise 14: Which points are located at the coordinates in the chart below?

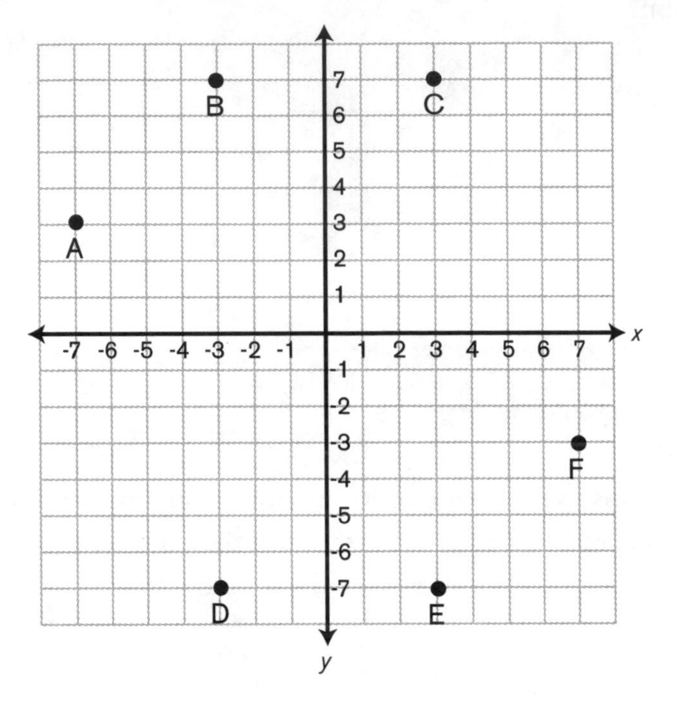

Ordered Pair	Letter
(–3, 7)	
(7, –3)	
(–3, –7)	
(–7, 3)	
(3, 7)	

Line AB below contains the points (2, 3) and (–3, –2). How would you calculate the equation of line AB?

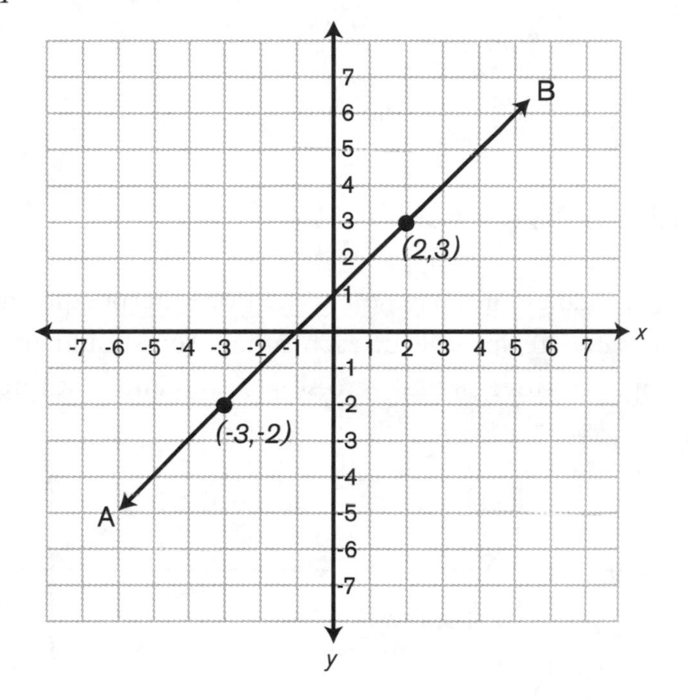

The equation for a line is:

$$y = mx + b$$

Here, ***m*** is the slope of the line ($\frac{\Delta y}{\Delta x}$) and ***b*** is the *y* intercept. You are given two points to work with, so first you will determine the slope. The slope is simply the change in the *y* values of the coordinates divided by the change in the *x* values of the coordinates.

$$m = \frac{\Delta y}{\Delta x} = \frac{y_2 - y_1}{x_2 - x_1}$$

$$= \frac{3 - -2}{2 - -3}$$

$$= \frac{3 + 2}{2 + 3}$$

$$= 1$$

Putting $m = 1$ into the equation $y = mx + b$, you get $y = x + b$. You can use one (x, y) pair to figure out what b is. Let's use the coordinate (2, 3) and stick the x and y values into the equation below:

$$y = x + b$$

$$3 = 2 + b$$

$$b = 1$$

So, your final equation is $y = x + 1$

Interesting Facts: Parallel lines have the same slope. Perpendicular lines have slopes that are negative reciprocals of each other. You should also be able to spot negative slopes, positive slopes, zero slopes, and lines with no slope.

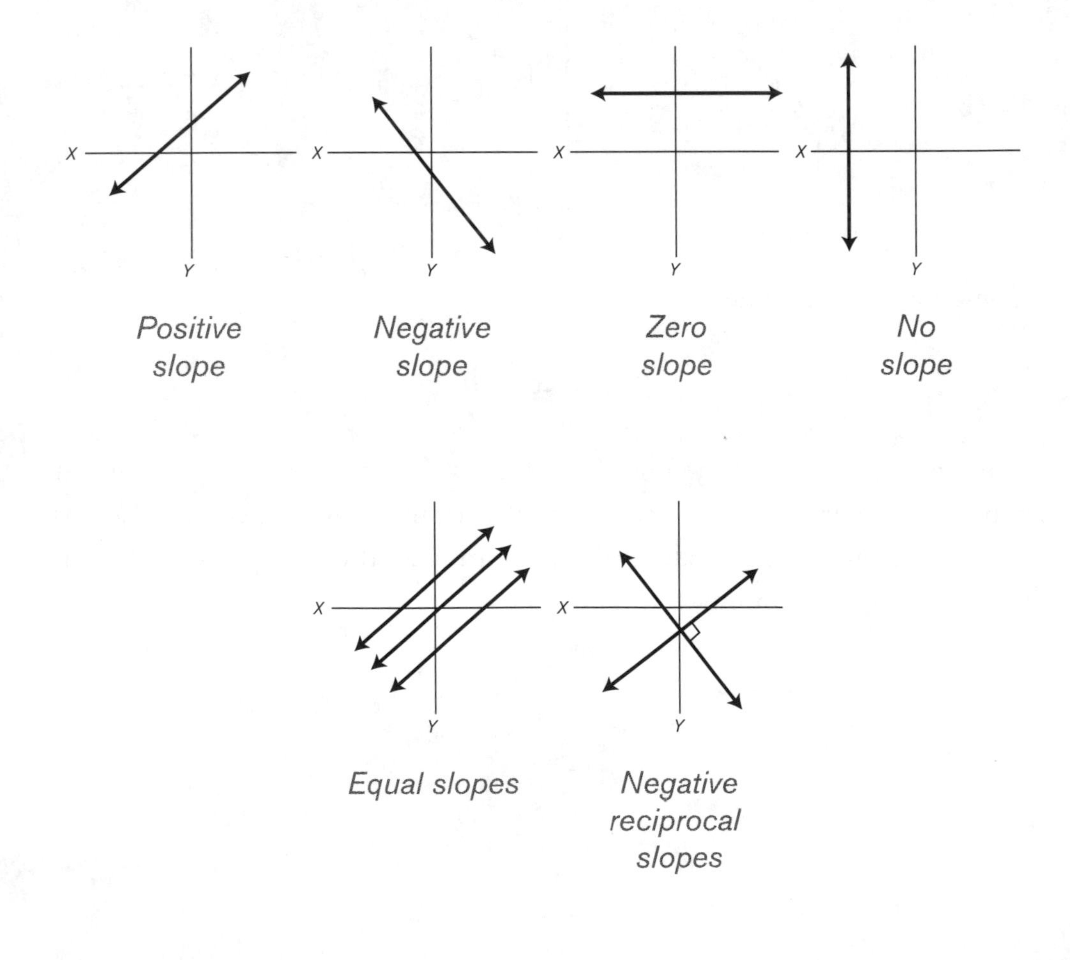

Exercise 15: What is the slope of the line below?

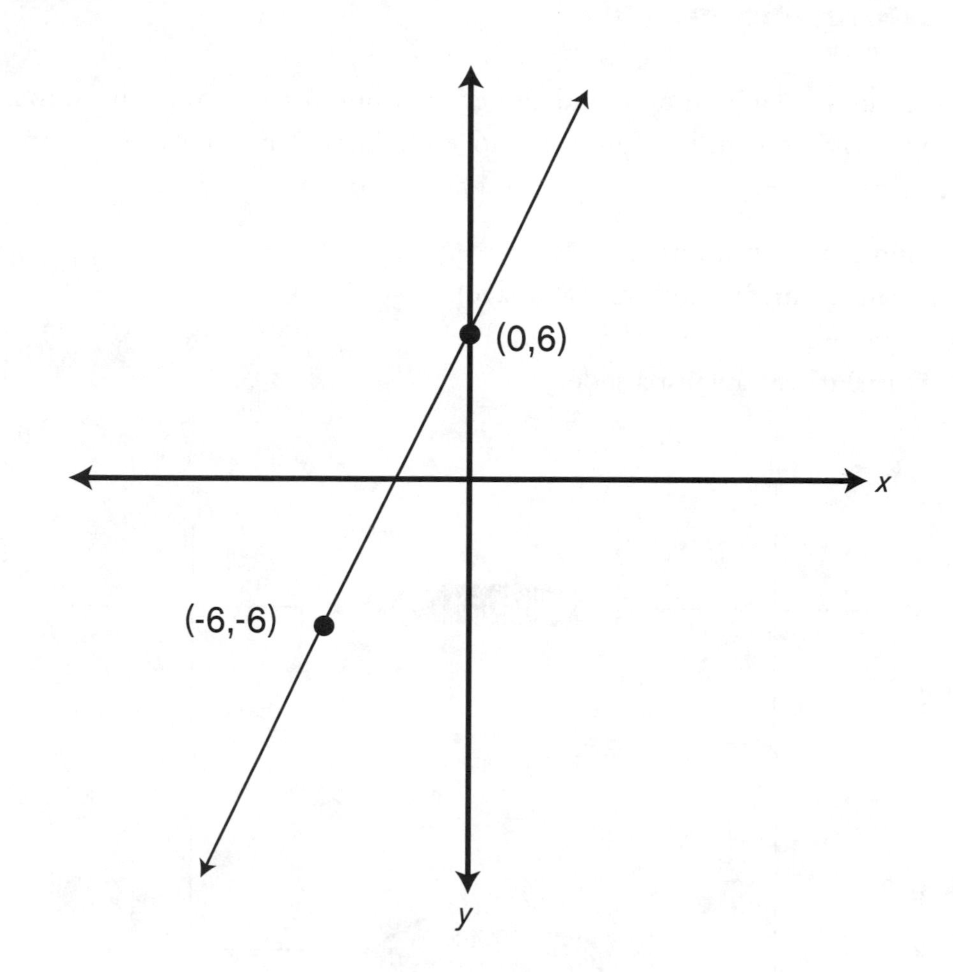

transformations

A given figure can be moved and altered in a number of ways. The way in which the old point of the figure are moved to "new" spots is called a **transformation**.

Common transformations are:

- A **reflection** is a flip (mirror-image).
- **Rotations** are turns.
- **Translations** are just a slide.

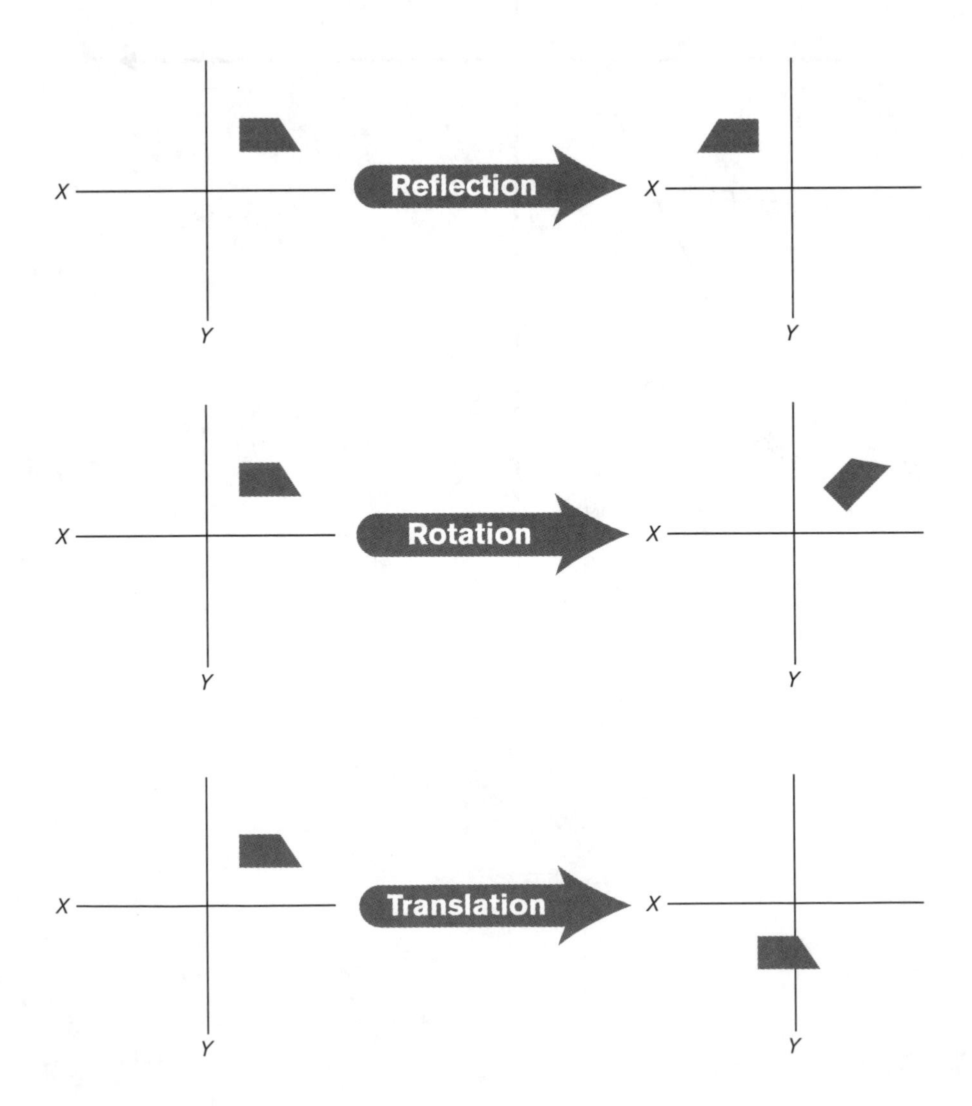

tessellations

A shape is said to **tessellate** if it can tile and cover a surface without any gaps. For example, the parallelogram below tessellates, because each "tile" locks into the next, leaving no gaps:

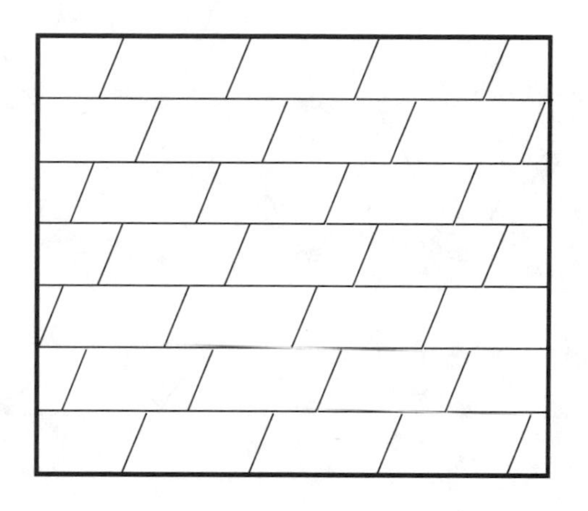

What shapes tessellate? Sometimes irregular shapes can tessellate too:

Tessellations can be used as a form of art. Here is an example of a more artsy type of tessellation:

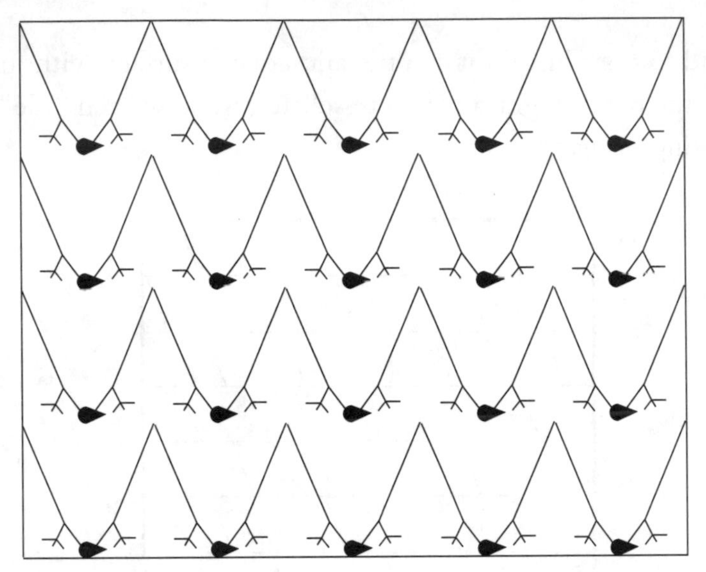

solutions to chapter exercises

Exercise 1:

6 yd	2 ft		8 in
+ 7 yd	1 ft		5 in
13 yd	3 ft		13 in
13 yd 1 yd			(12 in + 1 in)
14 yd	1 ft	+	1 in
14 yd	1 ft		1 in

Exercise 2:

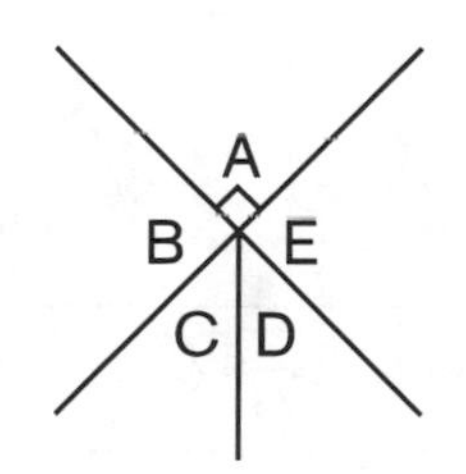

An example of . . .	Would be . . .
Vertical angles	∠B and ∠E
An acute angle	∠C and ∠D
A right angle	∠A, ∠B, and ∠E
Complementary angles	∠C and ∠D
Supplementary angles	∠A and ∠E ∠B and ∠A

Exercise 3:

Looking at the diagram, you can fill in 40° for the angle created when the diagonal line crosses ST.

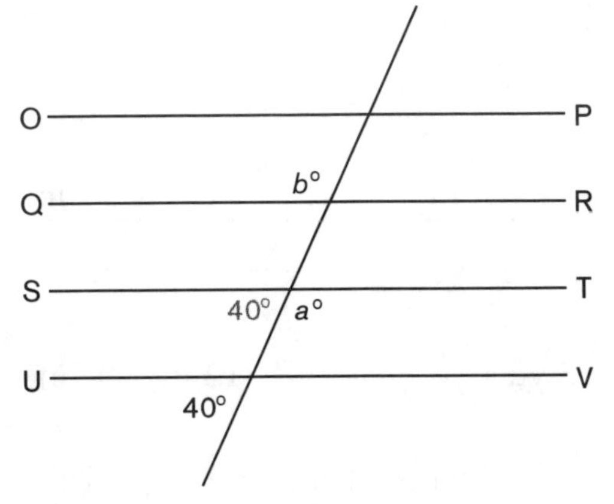

You can also fill 40° for the angle created when the diagonal crosses QR. This is because the diagonal line will create corresponding angles for all the parallel lines that it crosses. This newly labeled 40° angle and angle *b* make a straight line. Because straight lines are 180°, you know that angle *b* must be 140° (40° + 140° = 180°).

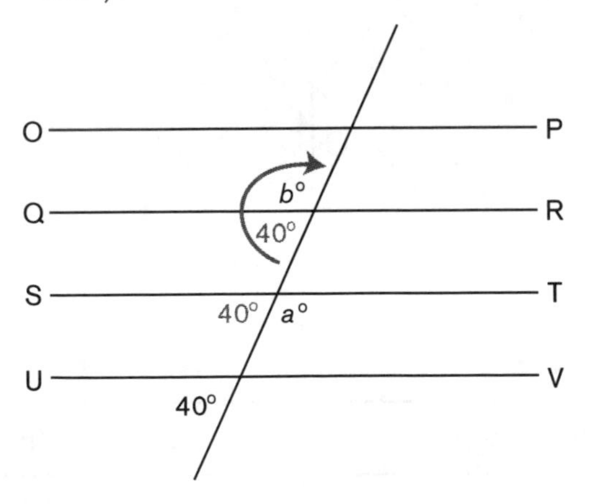

Exercise 4: Use **180°(*n* – 2)** with $n = 8$ (an octagon has 8 sides).
180(8 – 2) = 180(6) = 1080°.

Exercise 5: Given that $\overline{BC}$ is parallel to $\overline{DE}$, if the ratio of $\overline{AB}$:$\overline{BD}$ is 2:3, then you can calculate the ratio of $\overline{BC}$:$\overline{DE}$.

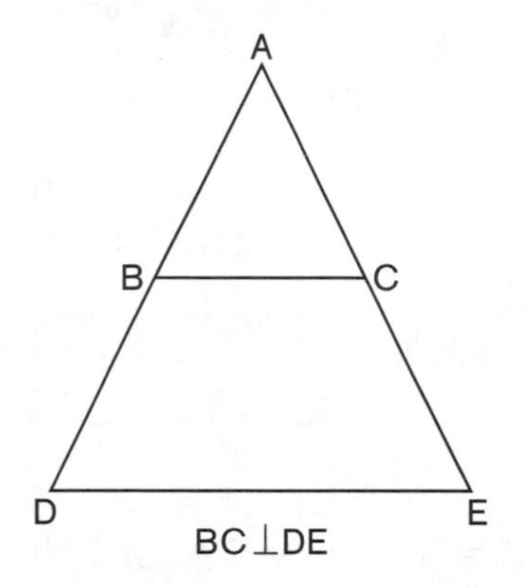

Notice that this figure contains similar triangles. Similar triangles are in proportion. Because AB:BD = 2:3, then you know that the triangles are in a 2:5 ratio.

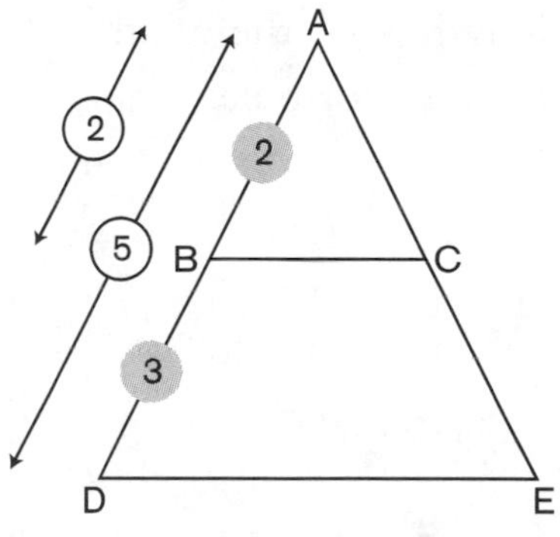

This means that the bases of these two triangles will also be in a 2:5 ratio.

Exercise 6: A regular pentagon has equal sides. So, you can write:

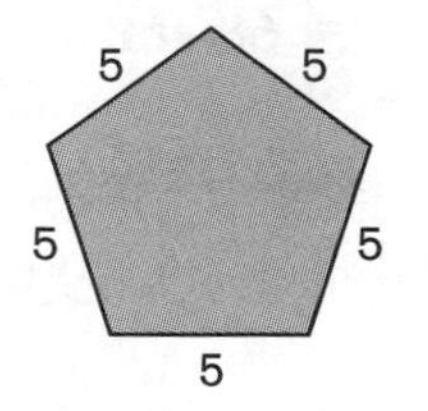

The perimeter, P = 5 + 5 + 5 + 5 + 5 = 25 mm.

Exercise 7: The area is currently

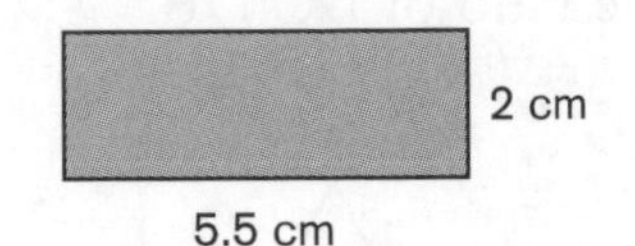

$A = L \times W = 5.5 \times 2 = 11\ cm^2$. If you double the length (the length is always longer than the width, so you double 5.5), the new length is 11 cm.

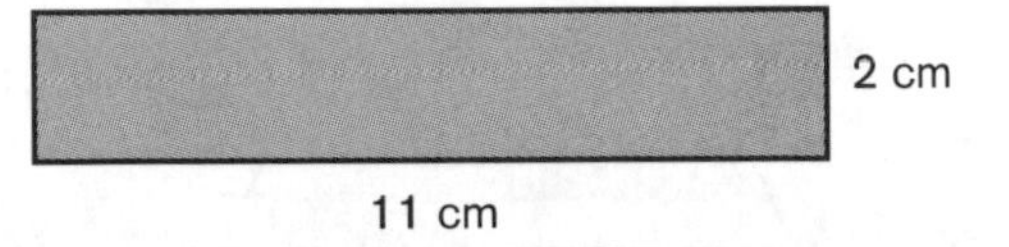

The new area is $L \times W = 11 \times 2 = 22\ cm^2$. How much greater than the original area is this? $22\ cm^2 - 11\ cm^2 = 11\ cm^2$.

Exercise 8: Use **Surface Area (S.A.) = 2(*lw*) + 2(*hw*) + 2(*lh*)**. From the figure you see that the length is 4″, the width is 1″, and the height is 6″. Substitute these values into the formula:

$$l = 4$$

$$w = 1$$

$$h = 6$$

$$\begin{aligned} \textbf{S.A.} &= \mathbf{2(lw) + 2(hw) + 2(lh)} \\ &= 2(4)(1) + 2(6)(1) + 2(4)(6) \\ &= 8 + 12 + 48 \\ &= 68\ cm^2 \end{aligned}$$

Exercise 9: If you use the formula $V = \pi r^2 h$, you can see that $r = 3$ and $h = 4$. Thus, the formula becomes $V = \pi(3)^2(4) = \pi(9)(4) = 36\pi$ feet^3.

Exercise 10: Using $a^2 + b^2 = c^2$, you know that the hypotenuse (or c) is 13, and one leg is 12. Substituting, you have: $a^2 + 12^2 = 13^2$. This means $a^2 + 144 = 169$, or $a^2 = 25$. Thus, $a = 5$.

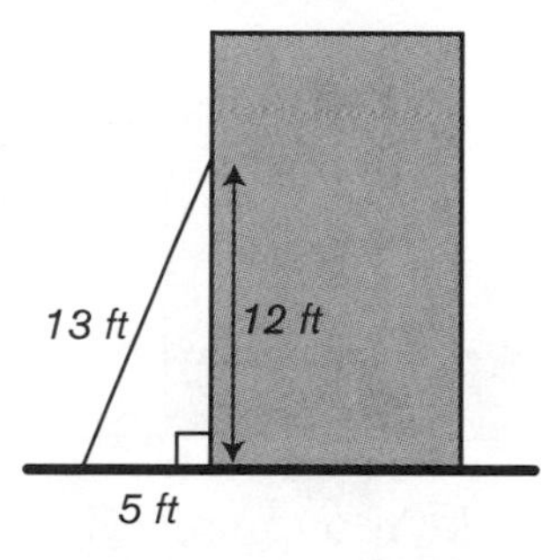

Exercise 11: A square is a rectangle because it has four sides and four right angles. A square is a rhombus because it has four equal sides and two sets of parallel sides.

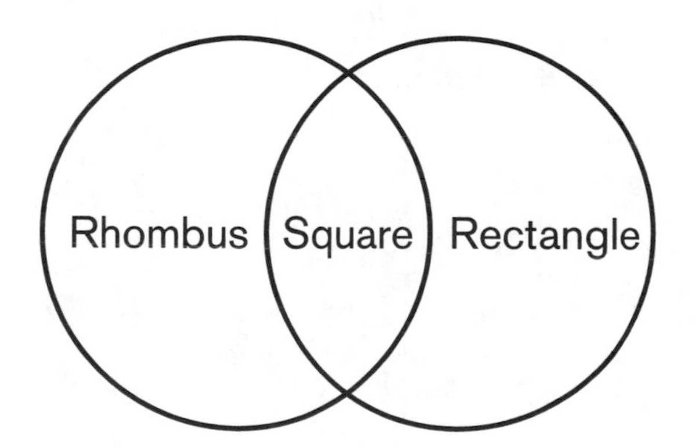

Exercise 12:

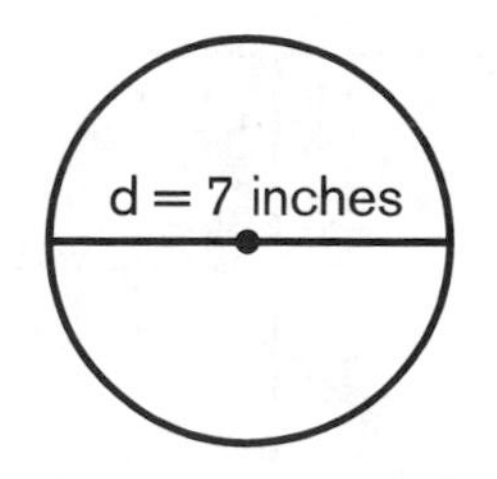

In the diagram, the 7 inches represent the *diameter*. The rolling wheel will have the distance around its outer edge in contact with the ground as it revolves. This question boils down to a *circumference* question. Using $C = \pi d$, with $\frac{22}{7}$ substituted in for π, this equation becomes: $C = \frac{22}{7}(7) = 22$ inches.

If the wheel revolves 20 times, you multiply 22×20 to get the number of inches traveled. This value, 440 in, can then be converted into feet: 440 in $\times$ $\frac{1 \text{ ft}}{12 \text{ in}} = 36\frac{2}{3}$ ft, choice **a**.

Exercise 13:

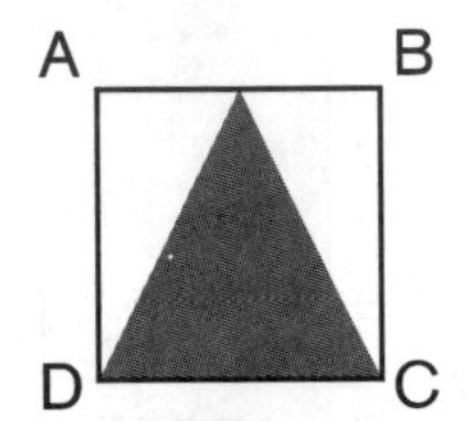

Use the formula $A = \frac{1}{2}\, b \times h$. You can see that the height and the base are equal to 16—the length of the side of the square.

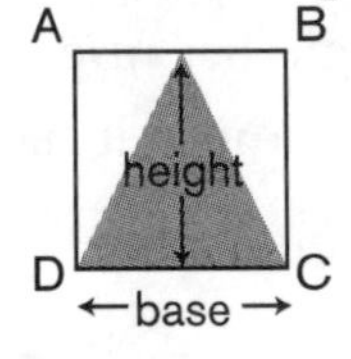

$A = \frac{1}{2}(16)(16) = 8 \times 16 = 128$ units^2.

Exercise 14:

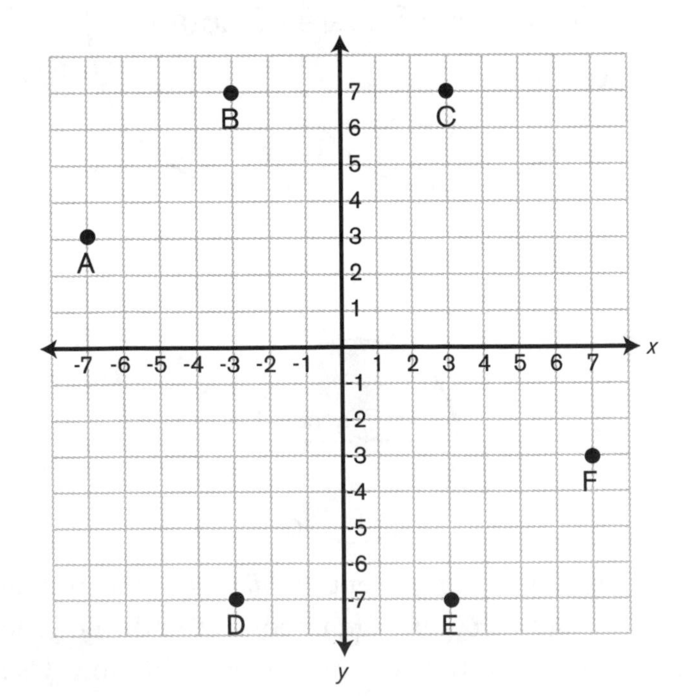

Ordered Pair	Letter
(–3, 7)	B
(7, –3)	F
(–3, –7)	D
(–7, 3)	A
(3, 7)	C

Exercise 15: Use $\frac{\Delta y}{\Delta x} = \frac{6 - (-6)}{0 - (-6)} = \frac{6 + {}^{+}6}{0 + {}^{+}6} = \frac{12}{6} = 2$

chapter
seven

Probability and Statistics

mean

Mean is another way of saying **average**. Averages are used to *typify* a group of numbers. To find the average, you total up all the values and then divide by the number of values.

$$\text{mean} = \frac{\text{sum of all values}}{\text{\# of values}}$$

On the following page, you see that Ann has \$6, Bob has \$4, and Carl has \$5. By simply shifting a coin over, you can see that the average of all three values would be \$5:

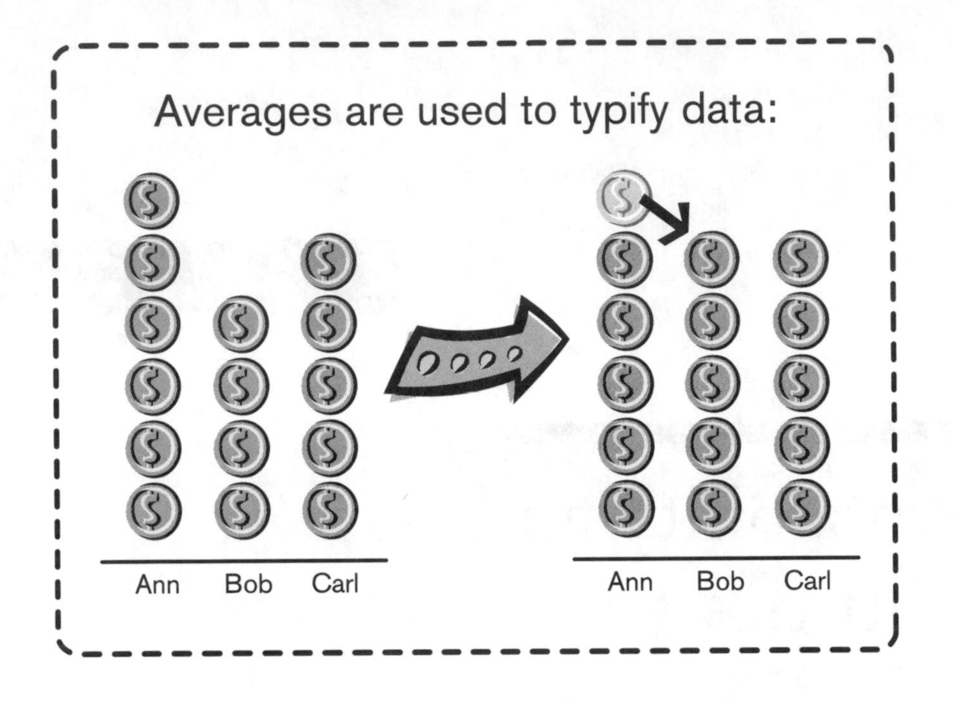

If you used the formula instead, you would total all three values, and divide by 3:

$$\frac{6 + 4 + 5}{3} = \frac{15}{3} = 5$$

Suppose you bought 5 CDs. The average price per CD came out to $13. If you remember that three CDs cost $12, and the fourth cost $15, what was the price of the fifth CD?

Here, use the formula for calculating the average, or the mean:

$$\text{mean} = \frac{\text{sum of all values}}{\text{\# of values}}$$

You know that the average is $13 and that 5 CDs were purchased. You have $13 = \frac{\text{(sum of all values)}}{5}$. Multiplying both sides by 5, you get 65 = sum of all CDs. Since you know that you bought 3 CDs at $12 each, a fourth for $15 and you need to figure out what the fifth CD cost. Put a "?" in for the fifth CD.

$$65 = \text{sum of all CDs}$$

$$65 = 12 + 12 + 12 + 15 + ?$$

$$65 = 36 + 15 + ?$$

$$65 = 51 + ?$$

$$14 = ?$$

This means that the fifth CD cost $14.

Exercise 1: Devin went to the bookstore and bought four books at an average price of $18.95. If three of the books sold for $25.25, $14.95, and $19.95, what was the cost of the fourth book?

median

When considering a list of values in order (from smallest to largest), the **median** is the middle value. If there are two "middle" values, then you just take their average.

Let's find the median of 2, 8, 3, 4, 7, 6, and 6. First you put these numbers in order:

2 3 4 6 6 7 8

Next, circle the middle number:

2 3 4 (**6**) 6 7 8

Next, let's find the median of: 2 3 4 4 6 6 7 8.

The numbers are already listed in order so you do not have to worry about arranging them. Notice that this list of numbers has two middle terms:

2 3 4 (**4 6**) 6 7 8

You take the average of these two numbers to find the median.

$$\frac{4+6}{2} = \frac{10}{2}$$
$$= 5.$$

Exercise 2: The quiz scores for seven students are listed below:

12 10 14 8 7 3 13

What is the median score?

mode

In a list of values, the **mode** is the number that occurs the most. If two numbers occur "the most," then you have two modes. This is called "bimodal."

Find the mode of the following numbers:

35 52 7 23 51 52 18 32

In this series of numbers: 35 52 17 23 51 52 18 32, you see that 52 appears twice.

35 **52** 17 23 51 **52** 18 32

Thus, the mode is 52.

Exercise 3: Find the mode of: 2 3 4 4 6 6 7 8.

probability

Probabilities help us predict the likelihood of events. For example, probability helps us predict the weather, the chance of rolling a certain number on a die, or the chance of picking a particular card out of a deck of cards.

In order to calculate the likelihood of a certain event, you just place the number of favorable outcomes over the number of total possible outcomes:

$$\text{Probability} = \frac{\text{\# favorable outcomes}}{\text{\# total outcomes}}$$

Suppose you have 3 blue socks, 5 purple socks, and 2 green socks in your dresser drawer. If you reach in and grab a sock, what is the probability that it will be green?

Your sock drawer:

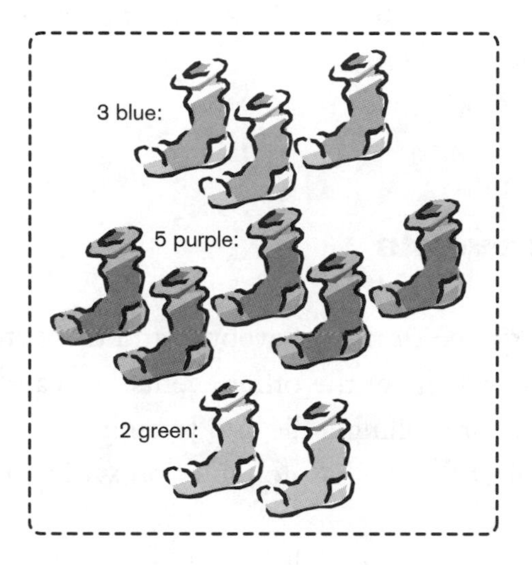

Now, a "favorable outcome" would be equivalent to a green sock. How many favorable outcomes are there?

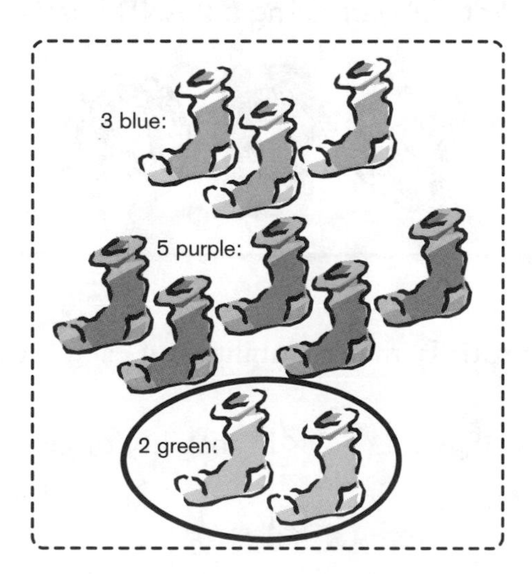

There are two green socks, and thus two favorable outcomes. The total number of outcomes is 3 + 5 + 2, or 10.

$$\text{Probability} = \frac{\#\text{ favorable outcomes}}{\#\text{ total outcomes}}$$

$$= \frac{2}{10}$$

$$= \frac{1}{5}$$

Exercise 4: Margaret tosses a six-sided die. What is the probability that the number she rolls is greater than 4?

independent events

If you are considering two or more "events," and the outcome of each event DOES NOT depend on any of the other events, we call this *independent*. For example, suppose you are rolling a die and tossing a coin. Whatever number comes up on the roll of the die has no effect on whether the coin comes up heads or tails.

Let's look at the probability of rolling a 2 and flipping "heads" on a coin.

Independent!

To solve, you just **multiply** the probabilities of each independent event:

Probability of getting a 2 × Probability of getting "heads"

$$\frac{1}{6} \times \frac{1}{2} = \frac{1}{12}$$

Even rolling the same die twice entails independent events. What is the probability of getting a 2 and then a 5?

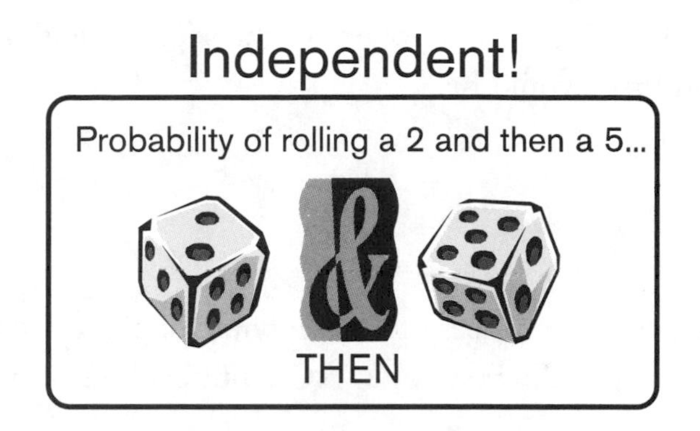

Probability of getting a 2 × Probability of getting a 5

$$\frac{1}{6} \times \frac{1}{6} = \frac{1}{36}$$

dependent events

As you probably have figured out, dependent events depend on one another. For example, if you took a coin out of a bag, stuck it into your pocket, and then reached in for a new coin, your chances have changed.

If a second event depends on a prior event, then when you describe the chances of it happening, you do so by saying, "The chances of event number two occurring GIVEN event number 1 is . . . "

For example, below you have 3 tokens and 2 dimes. Let's calculate the chances of choosing 2 tokens in a row, without replacement.

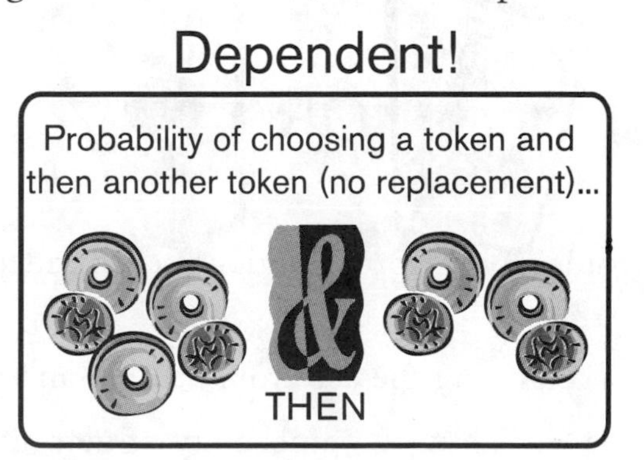

When the first coin is drawn, you have a $\frac{3}{5}$ chance of getting a token.

If you do get a token the first time, the chance of getting a token the second time would be $\frac{2}{4}$ or $\frac{1}{2}$.

Thus, the chances of getting a token on the second drawing, given that a token was drawn first, would be:

$$\frac{3}{5} \times \frac{1}{2} = \frac{3}{10}.$$

Exercise 5: What are the chances of randomly picking a red lollipop followed by a yellow lollipop if the first pop is NOT put back into the bag?

probability of one event or another

To calculate the probability of one event or another happening, you simply add up the two probabilities. For example, what is the probability of choosing an ace or a king from a deck of 52 cards?

Well, the chances of getting the ace would be $\frac{4}{52}$, and the chances of getting a king would be $\frac{4}{52}$. Thus, the probability of getting an ace OR a king would be $\frac{4}{52} + \frac{4}{52} = \frac{8}{52} = \frac{2}{13}$.

experimental probability

In theory, if you toss a die, you know that there is a 1 in 6 chance of getting a 2. But sometimes when you conduct dice-tossing experiments, you get statistics that do not reflect the theoretical probability. When analyzing results of such experiments, you should consider that maybe the die wasn't tossed enough times.

counting principle

If there are two events, and there are a ways that the first event can occur and b ways that the second event can occur, then there are $a \cdot b$ ways that these events can occur together. This is called the **counting principle**.

Let's say that you can buy white or dark chocolate hearts and they come in small, medium, or large sizes. Our total choices will equal *color of chocolate* × *size* = 2 × 3 = 6. You can see this below:

Exercise 6: At the diner, a kid's meal comes with a hamburger or veggie burger, fries or onion rings, and an ice cream, pudding, or shake. How many different kids' meals can be made?

permutations

The term *permutation* is just a fancy way of saying **arrangements**. When calculating a permutation, you are figuring out the number of possible arrangements. Just remember to use a **p**ermutation when **p**recise order matters.

__P__ermutations are for __P__recise Order!

What do we mean by "precise order?" Well, suppose you took the letters A, B, C, and D and you wanted to see how many different arrangements you could make.

Here, the precise order of the letters matters. Thus, ABCD ≠ BCAD ≠ CABD, and so forth.

To calculate a permutation for arranging ***a*** things in all into groups of ***c*** things chosen, you use this formula:

$$_{\text{ALL}}\mathbf{P}_{\text{CHOSEN}} = {}_{a}\mathbf{P}_{c}$$

Which means: Start taking the factorial of **a**, and stop after **c** spots.

A popular mnemonic to remember this is:

A Piece of Cheese

In our example you have ALL, or $a = 4$ (the four letters), and you are choosing all four, so $c = 4$ as well.

$$_{a}\mathbf{P}_{c} = {}_{4}\mathbf{P}_{4}$$

So you will take the factorial of 4 and stop after 4 spots:

$$\underset{\text{spot 1}}{\_\_\_} \times \underset{\text{spot 2}}{\_\_\_} \times \underset{\text{spot 3}}{\_\_\_} \times \underset{\text{spot 4}}{\_\_\_}$$

This is simply:

$$\underset{\text{spot 1}}{\underline{4}} \times \underset{\text{spot 2}}{\underline{3}} \times \underset{\text{spot 3}}{\underline{2}} \times \underset{\text{spot 4}}{\underline{1}}$$

$$= 24$$

Here are the 24 possible arrangements:

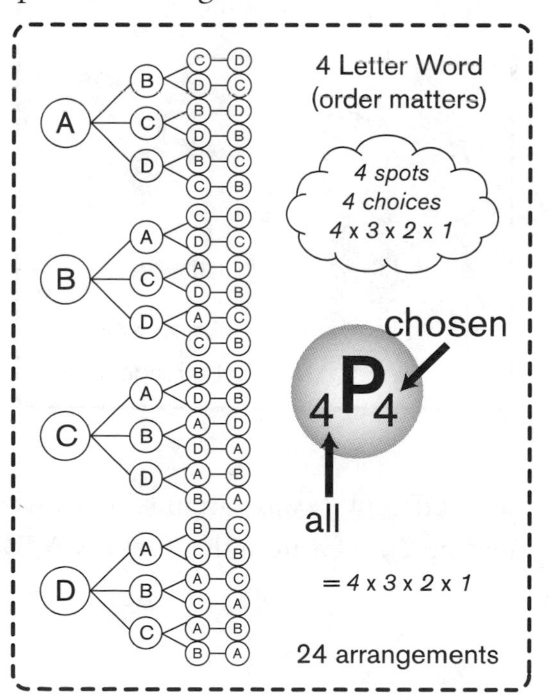

How many different "word" arrangements would be possible if you were only choosing 3 letters from the 4 letters A, B, C, D?

In this example, *a* is still 4 (the four letters). But now you are choosing 3, so CHOSEN, or $c = 3$ this time.

$$_{a}\mathbf{P}_{c} = {_{4}\mathbf{P}_{3}}$$

This is simply:

$$\underset{\text{spot 1}}{\underline{4}} \times \underset{\text{spot 2}}{\underline{3}} \times \underset{\text{spot 3}}{\underline{2}}$$

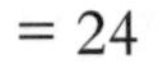

$= 24$

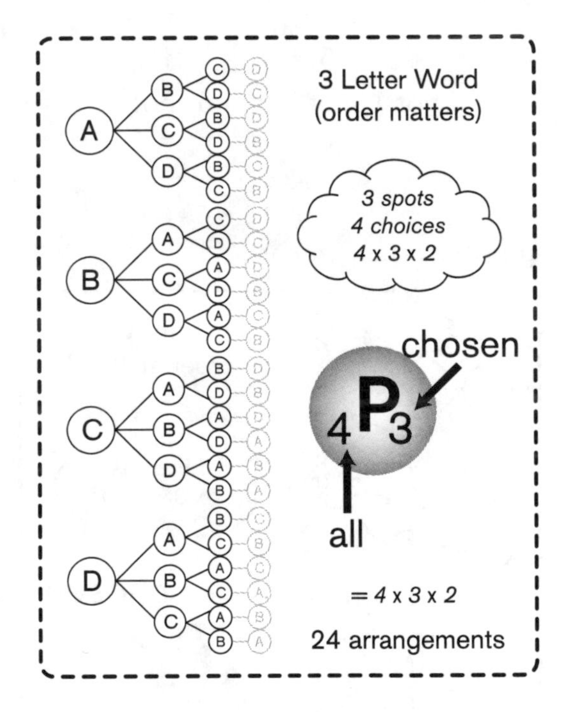

Exercise 7: How many different "word" arrangements would be possible if you were only choosing 2 letters from the 4 letters A, B, C, D?

combinations

When you are choosing a group (or cluster) of items and the order does not matter, you are calculating a combination.

Combinations are for Clusters where order does not matter!

Consider this scenario:

Geri has 3 different types of stickers: stars, hearts, and clovers. She will put 2 stickers in each party bag. How many different combinations are possible?

The possible arrangements (permutation) would be ${}_3P_2 = 3 \times 2 = 6$, but notice that because the stickers are just going in a bag, the order does not matter. This means that some of the sets listed are equivalent.

SHC = SCH

HCS = HSC

CHS = CSH

Thus, there are only 3 combinations. In this case, it was easy to just count them, but what if you had 100 stickers to choose from and you were putting 2 in each envelope? That's where using the combination formula comes in handy. First, let's look at the formula when you have 3 stickers and you are picking 2:

$$ {}_{ALL}C_{CHOSEN} = \frac{{}_aP_c}{c!} $$

Here, a good mnemonic to use is:

A Piece of Cheese over Crackers!

$$_{a}P_{c} \div c!$$

$$_{ALL}C_{CHOSEN} = \frac{_{a}P_{c}}{c!}$$

becomes:

$$_{3}C_{2} = \frac{_{3}P_{2}}{2!}$$

$$= \frac{(3 \times 2)}{(2 \times 1)}$$

$$= \frac{6}{2}$$

$$= 3$$

Let's see why this formula works. Suppose that 4 dancers are auditioning for a part in a performance, but there are only 3 spots open. Let's call the 4 dancers A, B, C, and D. When 3 dancers are picked, ABC = ACB = BAC = BCA = CAB = CBA. Calculating $_{4}P_{3}$ will give us too many results because ORDER DOES NOT MATTER in this case.

In the following diagram, notice that each possible trio is counted 6 times. For example, the shaded circles are all equivalent to A, B, and C getting the parts.

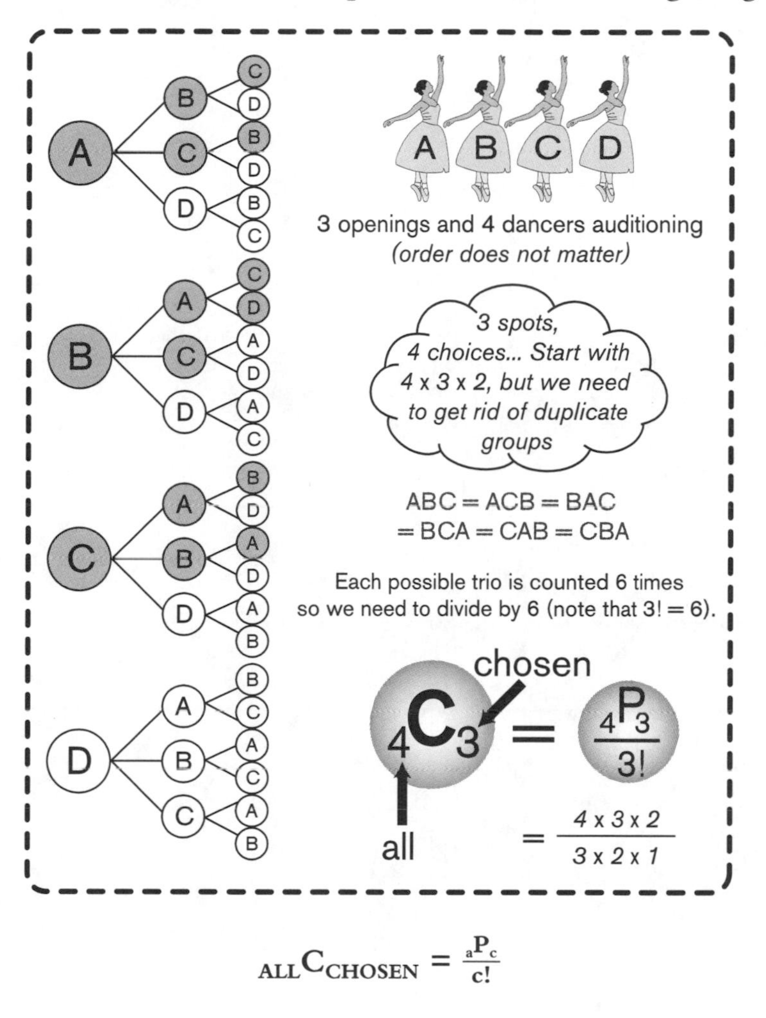

$$_{\text{ALL}}C_{\text{CHOSEN}} = \frac{_aP_c}{c!}$$

becomes:

$$_4C_3 = \frac{_4P_3}{3!}$$

$$= \frac{(4 \times 3 \times 2)}{(3 \times 2 \times 1)}$$

$$= \frac{24}{6}$$

= 4 possible trios

In order to come up with the correct number of groups, you simply divide by 3!, or 6.

Exercise 8: Denise wrote 5 names on a piece of paper, put them in a bag, and chose 3 names from the bag. How many different combinations are possible?

solutions to chapter exercises

Exercise 1: To find the average, you use the following formula:

$$mean = \frac{sum\ of\ all\ values}{\#\ of\ values}$$

Here you know the average is \$18.95, and that the number of values = 4 (there are 4 books). Substituting, you get:

$$\$18.95 = \frac{sum\ of\ all\ values}{4}$$

Cross-multiplying, you get \$18.95 • 4 = *sum of all values*, which means \$75.80 = *sum of all values*. So, the 4 books totaled \$75.80, and you know the price of 3 of these books. Therefore, you can subtract off the price of the 3 books to find out the cost of the fourth: \$75.80 – \$25.25 – \$14.95 – \$19.95 = \$15.65.

Exercise 2: To find the median score, you first must list all of the scores in order:

3 7 8 10 12 13 14

The middle number will be your median:

3 7 8 (10) 12 13 14

Hence, 10 is the median.

Exercise 3: When you look at: 2 3 4 4 6 6 7 8, you notice that both 4 and 6 occur the most. Therefore the mode = 4 and 6.

Exercise 4: All 6 possible outcomes:

1, 2, 3, 4, 5, 6

Only 2 possibilities (5 and 6) are greater than 4:

1, 2, 3, 4, (5, 6)

$$\text{Probability} = \frac{\#\text{ favorable outcomes}}{\#\text{ total outcomes}}$$

$$= \frac{2}{6}$$

$$= \frac{1}{3}$$

Exercise 5: When picking for the first time, there is a 3 out of 8 chance that red will be drawn. Now, GIVEN the first pop drawn was, in fact, red, you will have a $\frac{1}{7}$ chance of picking yellow:

Thus, picking a red lollipop followed by a yellow lollipop $= \frac{3}{8} \times \frac{1}{7} = \frac{3}{56}$.

Exercise 6: There are 2 choices of burgers, 2 choices of side dishes, and 3 choices of desserts. Thus, the total number of choices is: $2 \times 2 \times 3 = 12$. You can see this below:

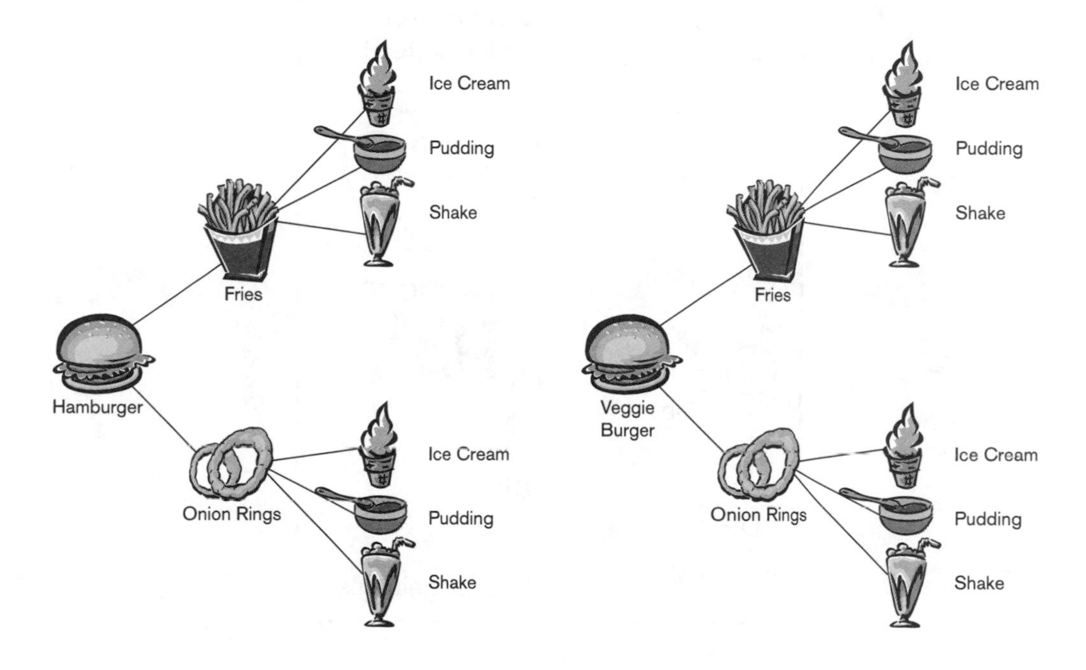

Exercise 7: Here a is still 4 (the four letters). But now you are choosing 2, so CHOSEN, or $c = 2$ this time.

$$_{a}\mathbf{P}_{c} = {}_{4}\mathbf{P}_{2}$$

So you will take the factorial of 4 and stop after 2 spots:

$$\underset{\text{spot 1}}{\underline{\quad}} \times \underset{\text{spot 2}}{\underline{\quad}}$$

This is simply:

$$\underset{\text{spot 1}}{\underline{4}} \times \underset{\text{spot 2}}{\underline{3}}$$

$$= 12$$

Here you see the 12 possibilities:

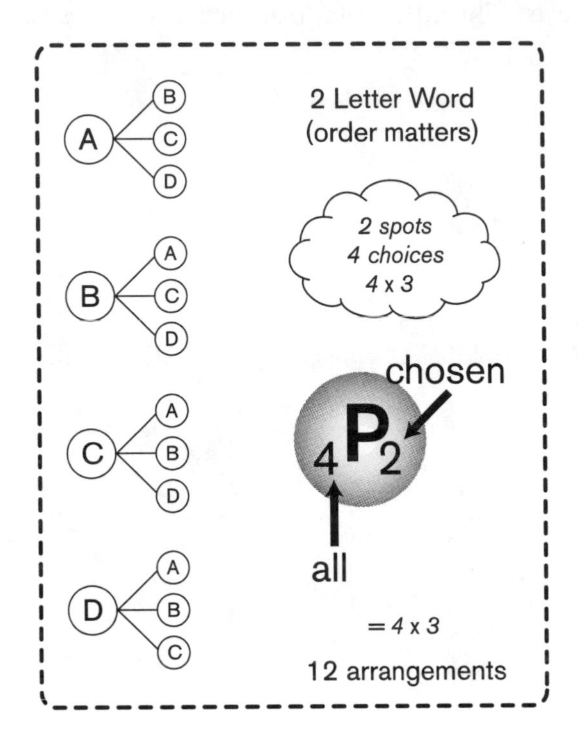

Exercise 8: Here, the order does not matter: If the five names are A, B, C, D, and E, picking ABC is the same as BCA. That is why you use the formula for combinations. All = 5 because there are 5 names, and c = 3 because she will choose 3. Thus, our formula

$$_{\text{ALL}}C_{\text{CHOSEN}} = \frac{_aP_c}{c!}$$

becomes:

$$_{\text{ALL}}C_{\text{CHOSEN}} = \frac{_5P_3}{3!}$$

On top, you start taking the factorial of 5, but you stop after 3 spots. On the bottom you take 3 factorial:

$$\frac{5 \times 4 \times 3}{3 \times 2 \times 1}$$

The two 3s cancel, so you have $\frac{5 \times 4}{2 \times 1}$, which is $\frac{20}{2}$, or 10.

chapter

eight

Tables and Charts

tables

Tables are used to organize information into columns and rows. Usually, a description of the data presented is located at the top of the table. Before looking to the particulars of the data, you should ask yourself, "What is this table telling me?" By focusing on what the table means to "tell you," you will be able to find the data that you need more easily, and analyze it accordingly.

For example, let's look at the following table:

Substance	Weight (lbs/ft^3)
Pumice	40
Saltpeter	75
Sand, dry	101
Sand, wet	120

What does this table tell you? It tells you the weight of different substances. Next, you will use this table to calculate the ratio of wet sand to pumice (in pounds per cubic feet).

Look at the chart:

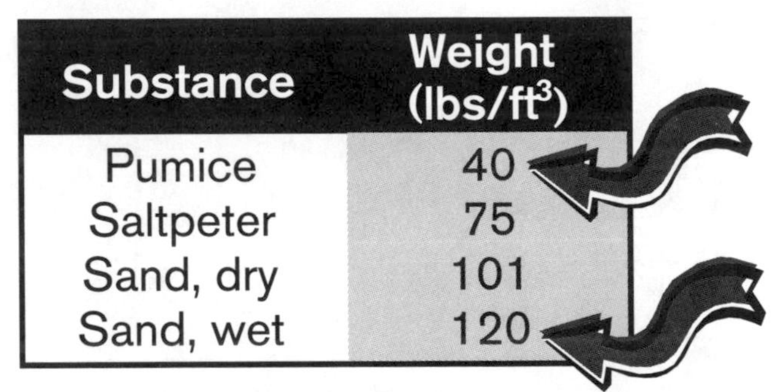

Substance	Weight (lbs/ft^3)
Pumice	40
Saltpeter	75
Sand, dry	101
Sand, wet	120

The weight in pounds per cubic feet of wet sand is 120.
The weight for pumice is 40. The ratio is then 120:40. This reduces to 3:1.

Exercise 1: A hospital is collecting donations to raise funds for a new wing. According to the table below, what was Nick's donation?

Donations	
Name	**Amount donated**
Kristi	$525
Ryan	$440
Nick	?
Toni	$615
Total	$2055

visual data presentation

There are many different ways that data can be presented in chart form. Let's first take a look at some common ways that this is accomplished.

common ways to present data

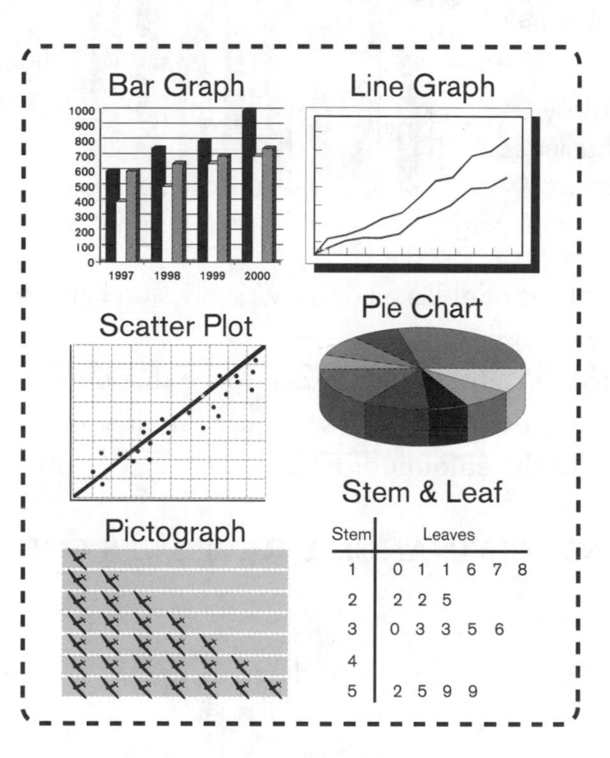

Now let's look at these different types of charts one by one.

bar graphs

Bar graphs are easy to read and are visually intuitive. Bar graphs can be used to present one type of data, or may contain different colored bars that allow for a side-by-side comparison of similar statistics. For example, on the following page is a triple bar graph. By looking at the key, you can see that this graph is simultaneously presenting data for East, West, and North. Always check your axes.

STEWART FINANCIAL REVENUES FOR 2001

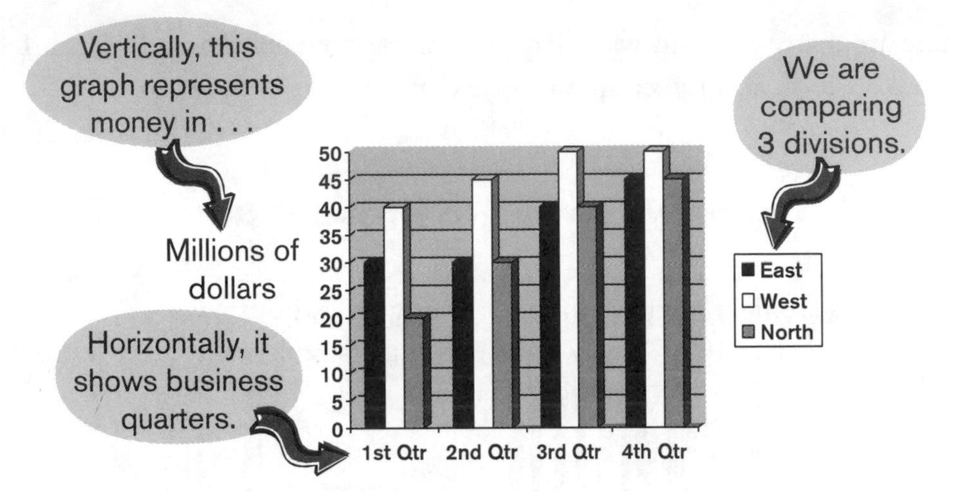

How many millions of dollars did the West division bring in throughout the first three quarters shown?

Use the key to the right of the diagram to see that the white bars represent the West revenues. Looking at the white bar for each of the first *three* quarters, you know that the amount of revenue is:

STEWART FINANCIAL REVENUES FOR 2001

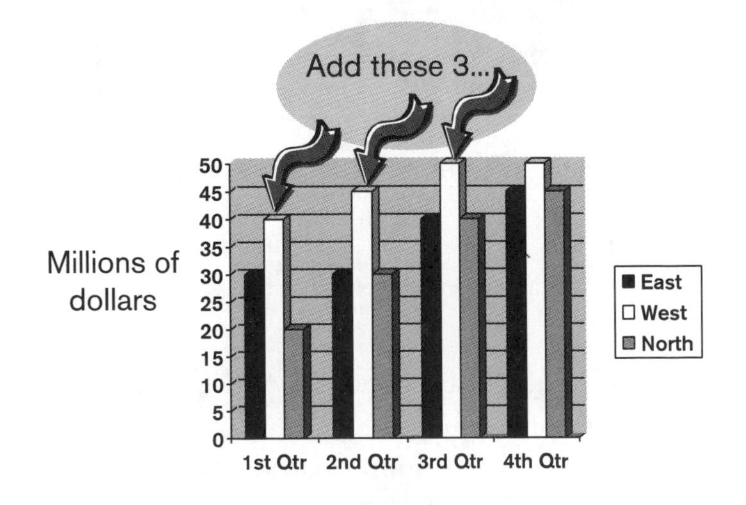

40 million + 45 million + 50 million = 135 million.

Exercise 2: Use the graph for Stewart Financial Revenues (on the last page) to answer the following question.

How much more did the East make in the third quarter than the North made in the first quarter?

line graphs

Line graphs aren't as pretty to look at as bar graphs, but they are easy to read. Again, different types of data may be presented simultaneously, as is the case below.

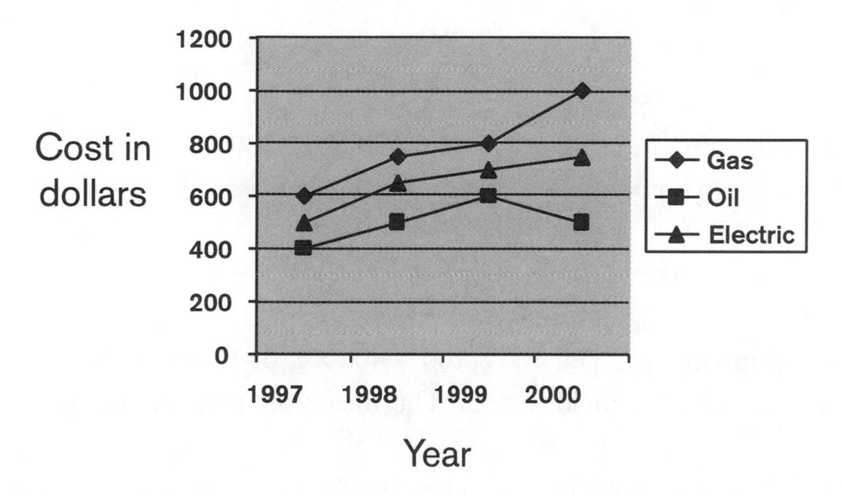

Look at the axes to see that this graph is telling us the monthly cost (in dollars) versus the year. Peek at the key to note that you are being given information on the money that goes toward Gas, Oil, and Electric.

Let's use this graph to calculate the percent increase in Oil cost from 1997 to 1998.

To calculate the percent increase, use this proportion:

$$\frac{change}{initial} = \frac{?}{100}$$

First, calculate the *change* by looking at the graph. Note that the lines with square points represent Oil.

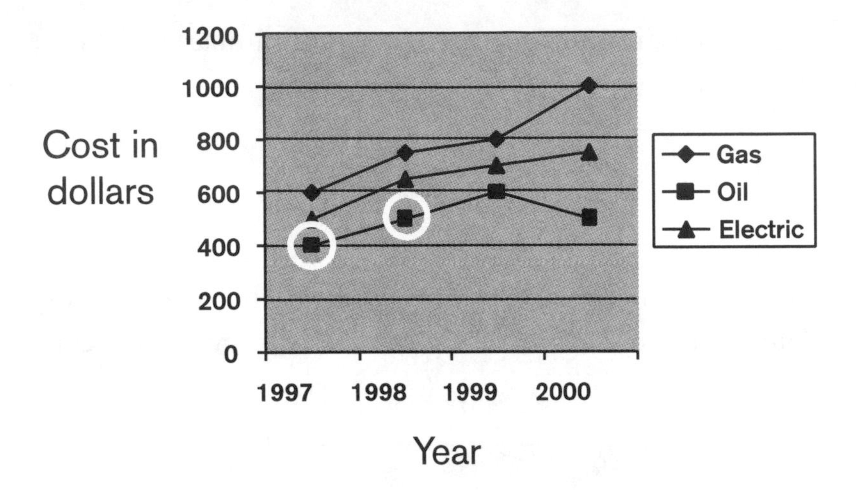

The cost rose from the *initial* $400 in 1997 to $500 in 1998. So, the *change* was $100. Thus, you get:

$$\frac{100}{400} = \frac{?}{100}$$

Cross-multiplying, you get: $100 \times 100 = 400 \times ?$, or $10{,}000 = 400 \times ?$, and dividing both sides by 400 yields $? = 25$. This means the percent increase was 25%.

Exercise 3: Use the graph below to best approximate the amount of bacteria (in grams) present on day 4.

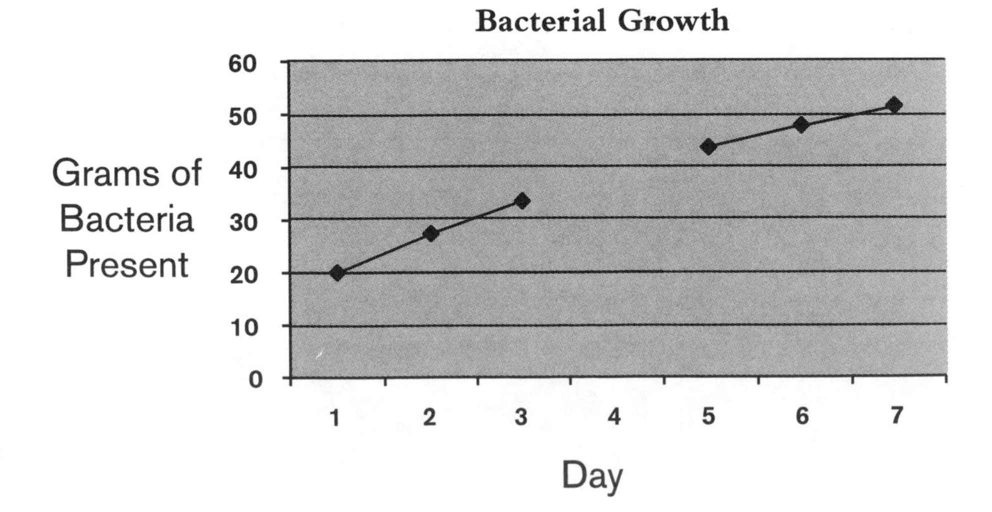

scatter plots

Scatter plots have points scattered all over the place. Usually you draw a "line of best fit" so that you can make sense of the data.

Here are the results from a survey at the Maynard School District. Notice the mess:

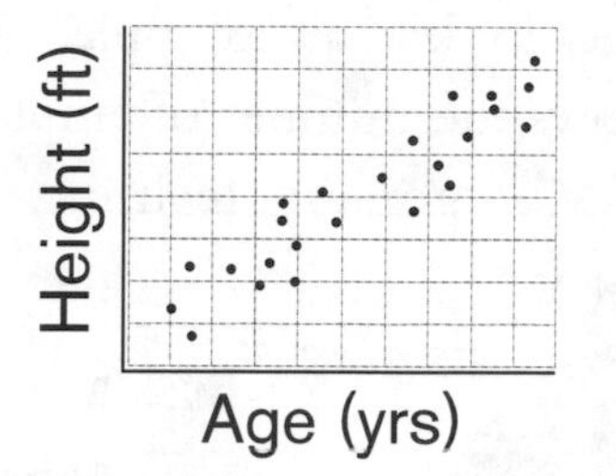

Now, draw a line that goes through the most points, and yet seems to have just the right amount of stray points scattered above and below:

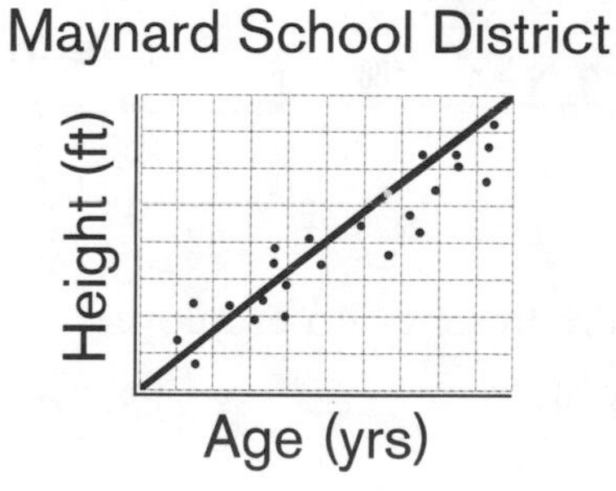

Exercise 4: Can you draw a line of best fit through the data points below?

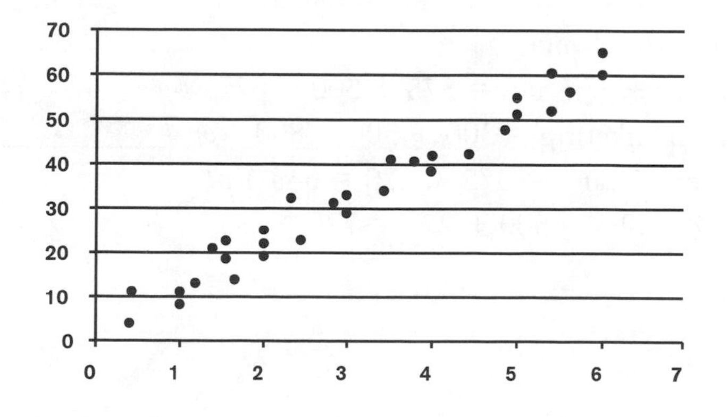

pie charts

Pie charts are a great way to visualize data. You can make a rough estimation regarding the different slices of the pie just by glancing at the chart. For example, if a slice is about a quarter of the pie, you can say, "Hey, that's about 25%." Sometimes these charts are two-dimensional, and sometimes they are three-dimensional. As always, you should approach a chart wondering, "What is this chart telling me?" Let's look at an example.

The pie graph below shows the number of employees working in each of the various departments of Montgomery Tech Inc.

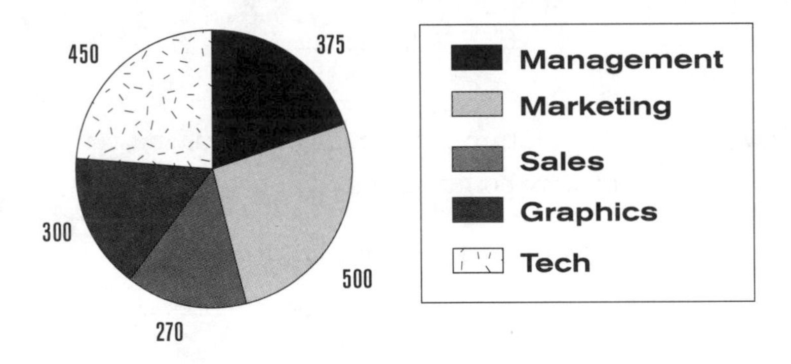

Which two departments represent 645 employees?

a. management + graphics
b. graphics + marketing
c. management + sales
d. marketing + sales

Let's go through the choices:

a. management + graphics = 375 + 300 = 675 *Nope!*
b. graphics + marketing = 300 + 500 = 800 *Nope!*
c. management + sales = 375 + 270 = 645 *Yep!*
d. marketing + sales = 500 + 270 = 770 *Nope!*

Choice **c** is correct.

Exercise 5: The student enrollment at Lafayette Technical Institute is given below in the form of a three-dimensional pie chart in which students are grouped according to their course of study. What is the ratio of programming students to multimedia students?

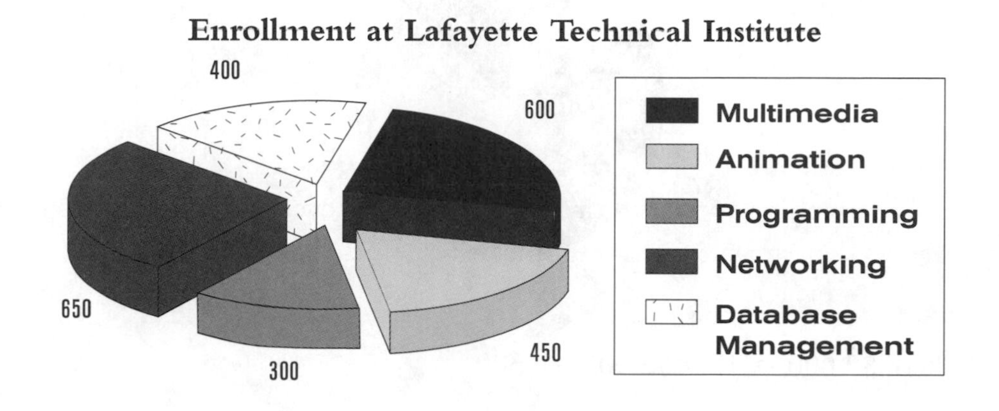

percent pie charts

Sometimes information is presented in terms of percent. For example, the pie chart below represents John's monthly expenses, which total $2,000. How much does John spend on his car each month?

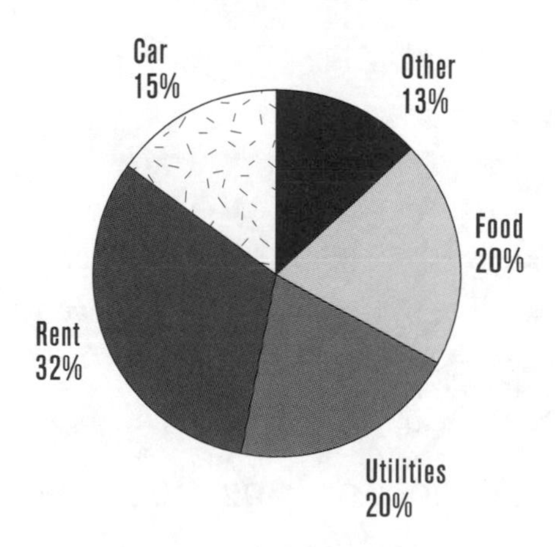

Look at the chart to see that Car represents 15% of the total $2,000 per month.

John's Monthly Expenses

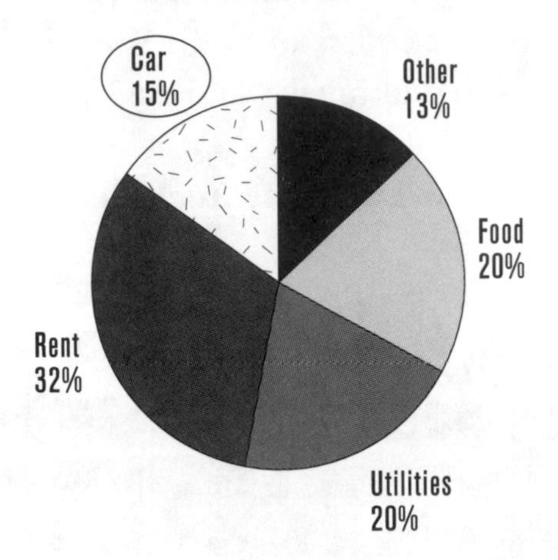

15% of \$2,000 = .15 × \$2,000 = \$300.

pictographs

Instead of using lines, bars, or chunks of pie to represent data, pictographs use pictures. You may see these types of graphs in newspapers and magazines. Here is an example of a pictograph:

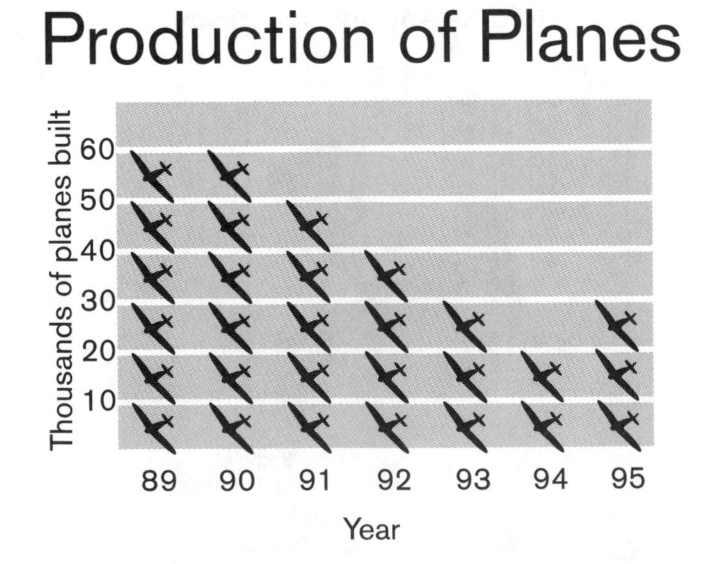

Exercise 6: While on a road trip, Kira made a pictograph of all the different vehicles that she saw. How many more motorbikes did she see than campers?

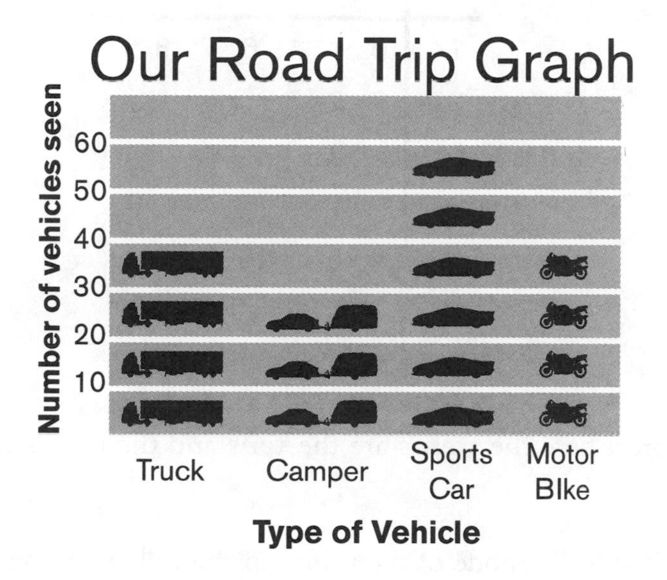

stem & leaf plots

Stem and leaf plots are the least intuitive of all the charts you have addressed so far. In other words, if you have never seen a stem and leaf plot before, it is hard to tell what one means. For example, here is a stem and leaf plot:

Stem	Leaves
1	0 1 1 6 7 8
2	2 2 5
3	0 3 3 5 6
4	
5	2 5 9 9

Key: 5 | 2 = 52

The key to understanding a stem and leaf plot is to look at the key:

Stem	Leaves
1	0 1 1 6 7 8
2	2 2 5
3	0 3 3 5 6
4	
5	2 5 9 9

Key: 5|2 = 52

In other words, here the **stems** are the **tens** and the **leaves** are the **ones**.

Exercise 7: What is the mode of the values in the following stem and leaf plot?

Stem	Leaves
1	0 1 1 6 7 8
2	2 2 5
3	0 3 3 3 6
4	2 5 6 7

Key: 2|5 = 25

solutions to chapter exercises

Exercise 1: First, let's add up all of the donations listed in the chart:

Donations	
Name	Amount donated
Kristi	$525
Ryan	$440
Nick	?
Toni	$615
Total	$2055

$525 + $440 + $615 = $1,580. Whatever Nick gave plus the $1,580 will yield $2,055. You can just subtract $1,580 from $2,055 to get Nick's donation: $2,055 – $1,580 = $475.

Exercise 2: For the third quarter, the black bar which represents the revenues for East goes up to 40 million dollars. For the first quarter, the gray bar, which represents the revenues for North, goes up to 20 million dollars.

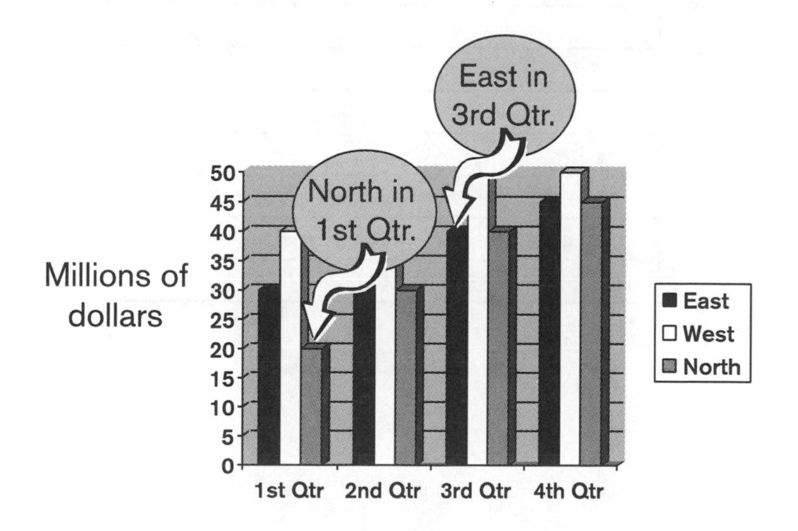

Thus, the East made 40 million – 20 million = 20 million dollars more than the North for the time periods cited.

Exercise 3: Fill in the missing line, based on your judgment:

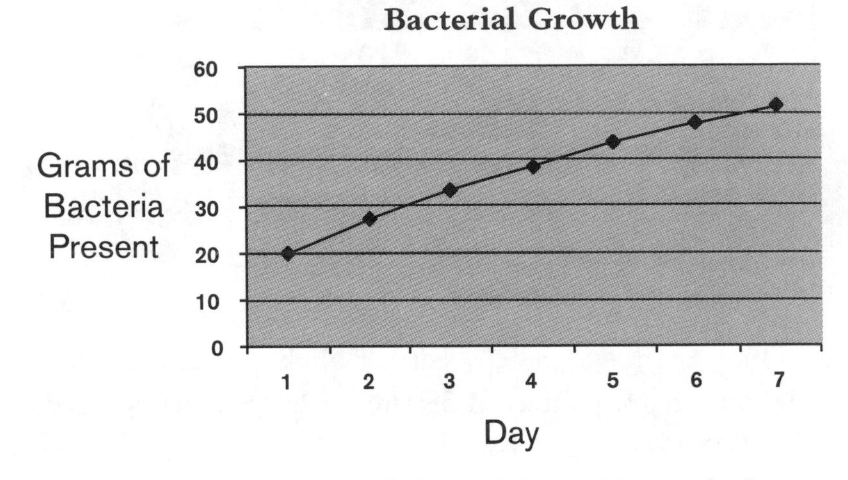

The best approximation is 38. If you said 37 or 39, you had a pretty good estimate.

Exercise 4:

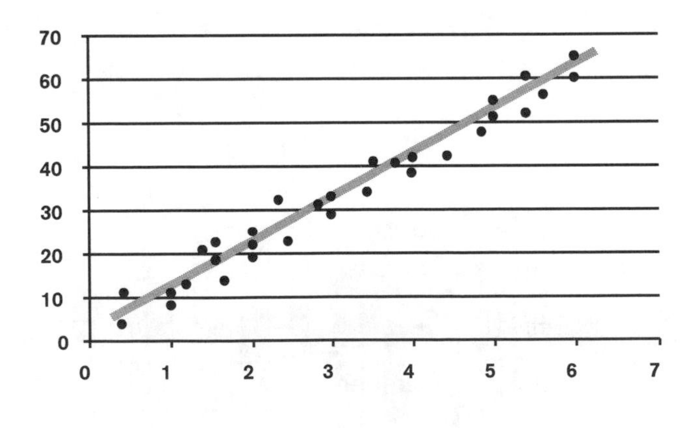

Exercise 5: Note that the black piece of the pie represents 300 programming students. The dark gray piece of the pie represents 600 multimedia students.

Enrollment at Lafayette Technical Institute

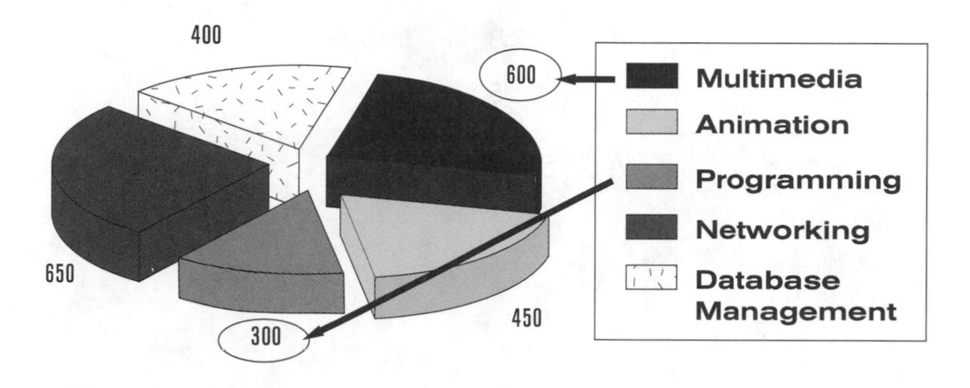

Thus, the ratio of programming students to multimedia students is 300:600. This reduces to 1:2.

Exercise 6: She saw 40 motorbikes and 30 campers:

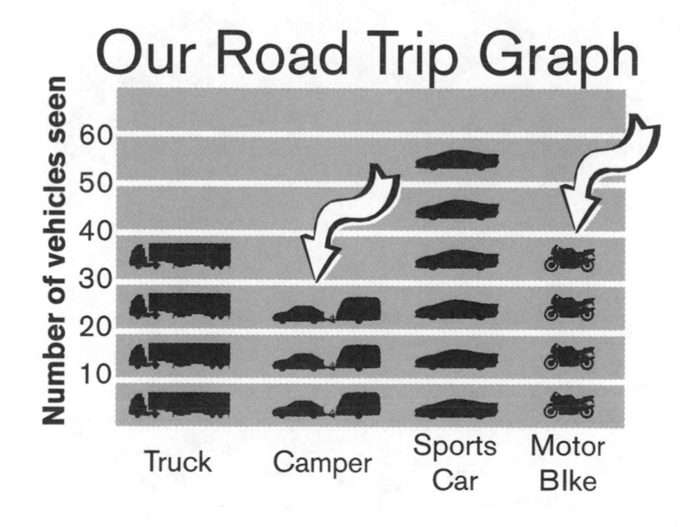

Thus, she saw 40 – 30 = 10 more motor bikes than campers.

Exercise 7: 3 occurs 3 times, and is thus the mode.

Stem	Leaves
1	0 1 1 6 7 8
2	2 2 5
3	0 3 3 3 6
4	2 5 6 7

Key: 2 | 5 = 25

Test Your Math Skills

the following test contains 35 multiple-choice questions. It is based on all of the topics presented in the preceding chapters. See how well you can do without peeking! To check your answers, turn to page 252.

1. In the figure below, Block A weighs 6 g and Block B weighs 12 g. If the fulcrum is perfectly balanced, how far away from the center is B?

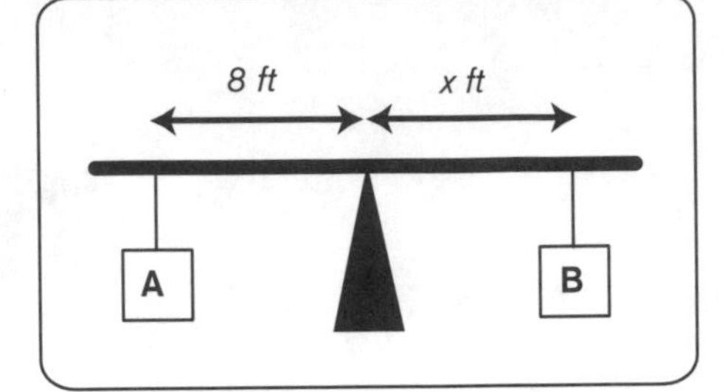

a. 10 ft
b. 8 ft
c. 6 ft
d. 4 ft

2. Ed is buying a jacket that originally cost $80. The tag on the sleeve reads "25% off." How much does the jacket cost?
a. $20
b. $40
c. $60
d. $70

3. What is the percent decrease in the number of trucks recalled from 1999 to 2000?

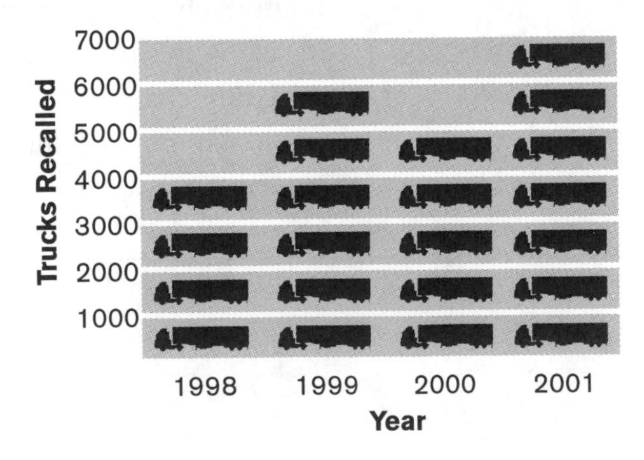

a. 20%
b. $16\frac{2}{3}$%
c. 15%
d. 10%

4. The table below shows the relationship between two variables: x and y. Which answer choice best represents this relationship?

x	y
0	0
1	–
2	5
3	–
4	17

a. $y = 2x + 1$
b. $y = 4x + 1$
c. $y = x^2 + 1$
d. $y = x^2$

5. Use the chart below to determine what the height of the water will be at 10 A.M., given the same rate of accumulation.

Time	Water Height
5 A.M.	23.02 cm
6 A.M.	23.23 cm
7 A.M.	23.44 cm

a. 23.65 cm
b. 24.02 cm
c. 24.07 cm
d. 26.21 cm

6. What fractional part of the figure below is shaded?

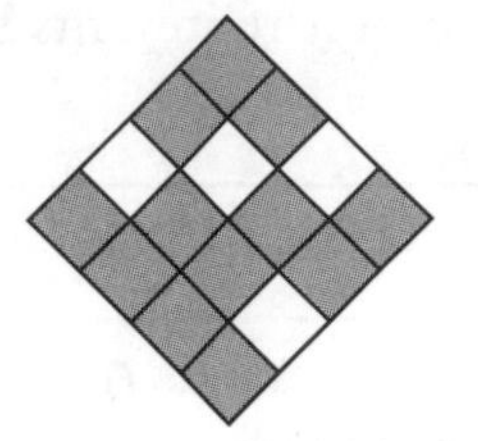

a. $\frac{1}{4}$

b. $\frac{3}{4}$

c. $\frac{1}{8}$

d. $\frac{3}{8}$

7. If BC is parallel to DE, and DB = 6, what is the value of AE?

a. 4

b. 6

c. 8

d. 10

8. Two baby chickens start running towards each other. One chick runs at a constant rate of three mph and the second chick runs at a constant rate of four mph. They meet after two hours. How far apart were they initially?

a. 14 miles

b. 16 miles

c. 60 miles

d. It cannot be determined from the information given.

9. What is the reciprocal of $\frac{1}{5}$?

a. $-\frac{1}{5}$

b. -5

c. $\frac{1}{5}$

d. 5

10. The solution to $x = \sqrt{10 + 1}$ can be classified as

a. a prime number.

b. a rational number.

c. an integer.

d. an irrational number.

11. 3.1×3^{10} when compared with 3.1×3^{8}

a. is 2 times as large.
b. is 3 times as large.
c. is 9 times as large.
d. is 3^3 times as large.

12. Given $-8x + 2 = 5x + 1$, what is the value of x?

a. -13
b. $\frac{1}{13}$
c. 13
d. $-\frac{1}{13}$

13. The pie chart below shows the percent sales for TrueTech. If sales totaled $270,000, how much money did the "Overseas" division bring in?

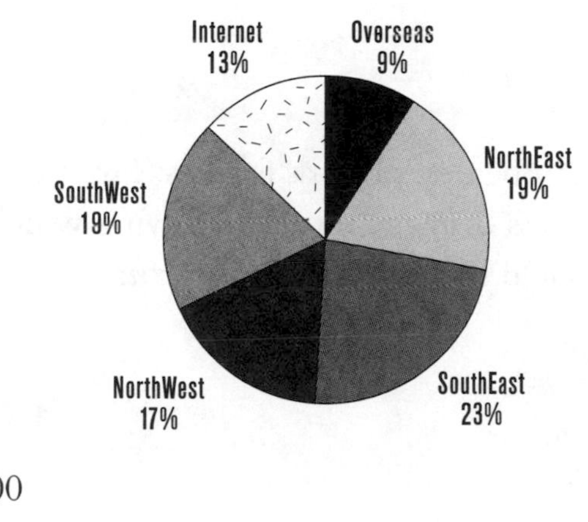

a. $243,000
b. $51,300
c. $24,300
d. $5,130

14. A six-sided die with sides numbered one through six is rolled. What is the probability that the number rolled is a multiple of three?

a. $\frac{1}{3}$
b. $\frac{1}{6}$
c. $\frac{2}{3}$
d. $\frac{3}{6}$

15. Five dancers are performing in a recital. Sammy will stand in the middle because she is the smallest. If the other four dancers will be arranged on either side of her, how many arrangements are possible?

a. 120

b. 24

c. 12

d. 6

16. Which of the following figures will tessellate?

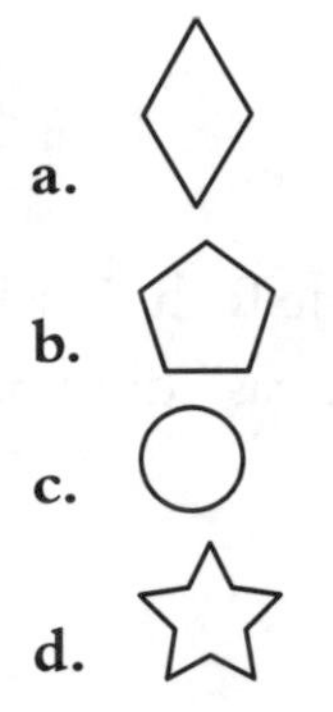

17. A number, n, can be described as follows: Five plus a number is multiplied by three less than the number. Solving which of the following equations will yield the correct values for n?

a. $5n \times 3n$

b. $n + 5 \times n - 3$

c. $(n + 5)(3 - n)$

d. $n^2 + 2n - 15$.

18. In the figure below, what is the height, h, of the telephone pole?

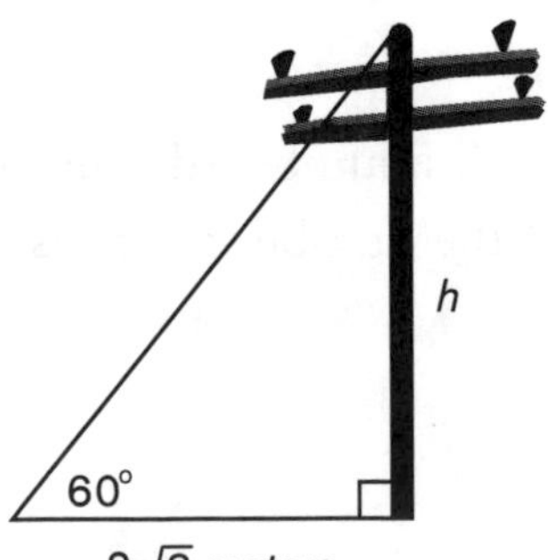

a. $18\sqrt{3}$
b. 6
c. $9\sqrt{3}$
d. 9

19. Use the formula I = PRT to answer the following question:

How long will $3,200 have to be invested at 8% to earn $768 in interest?

a. 1 year
b. 2 years
c. 3 years
d. 4 years

20. The expression $\frac{x^2 - 49}{x^2 + 6x - 7}$ is equivalent to:

a. $\frac{x - 7}{x - 1}$
b. $\frac{x + 7}{x + 1}$
c. $\frac{x - 7}{x + 1}$
d. $\frac{x + 7}{x - 3}$

21. Eight software developers will havc an opportunity to speak briefly at a computer show. Evan needs to list the developers in the order in which they will speak. How many different arrangements are possible?

a. 56
b. 336
c. 1,680
d. 40,320

22. If line segment AB is parallel to line segment CD, what is the value of x?

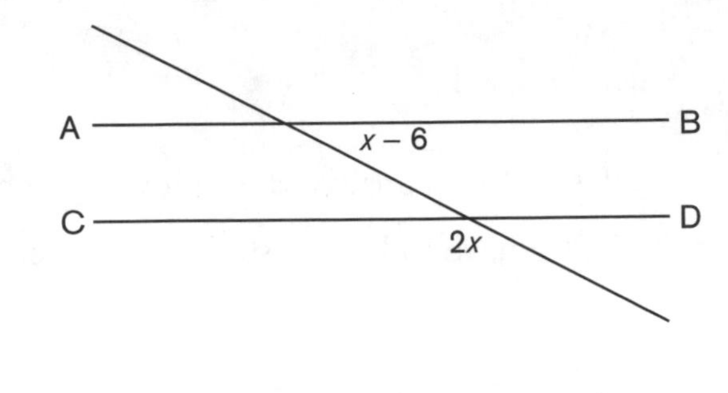

a. 58°
b. 62°
c. 56°
d. 60°

23. Use the chart below to determine the mean score of the people listed.

Name	Score
Alec	75
George	81
Felicia	93
Maria	77
Ashley	84

a. 82
b. 102
c. 76
d. 89

24. What is the perimeter of the rectangle shown below?

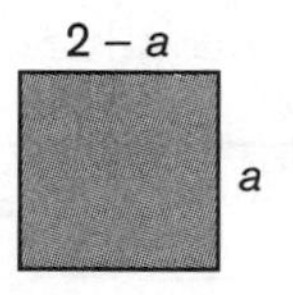

a. $(2 - a)^2$

b. $(2 - a)(a)2$

c. $4 + 8a$

d. 4

25. If $a = 2$, $b = -1$, and $c = \frac{1}{2}$, $\frac{2a - b + 5}{c}$ is equal to which of the following?

a. 5

b. 10

c. 20

d. 25

26. There were 504 candies in a jar. Between 8 o'clock and 9 o'clock, $\frac{1}{8}$ of the candies were given out. Between 9 o'clock and 10 o'clock, $\frac{2}{9}$ of the remaining candies were handed out. If in the following hour, $\frac{1}{7}$ of the remaining candies are distributed, how many candies will be left?

a. 336

b. 294

c. 188

d. 96

27. The chart below shows the distribution of elements, by percent, in the human body. What is the value of x?

Element	Percent by Weight
Carbon	18%
Hydrogen	10%
Oxygen	65%
Other Elements	x%
Total	**100 %**

a. 5
b. 7
c. 17
d. 25

28. Which statement is true regarding the following numbers:

12 14 15 19 20 22

a. The average is greater than the median.
b. The median is greater than the mean.
c. The mean equals the median.
d. The mean is greater than the average.

29. As a promotion, Marco's record company will hand out random CDs to a crowd. If the company brought along 300 hip-hop CDs, 500 alternative rock CDs, 200 easy listening CDs, and 400 country CDs, what is the probability that a given person will get an easy listening CD?

a. $\frac{1}{8}$
b. $\frac{1}{7}$
c. $\frac{1}{6}$
d. $\frac{1}{5}$

30. What is the log of 1×10^5?

a. 10^5

b. 10^{-5}

c. 5

d. -5

31. If the side of the square below is doubled, what is the effect on its area?

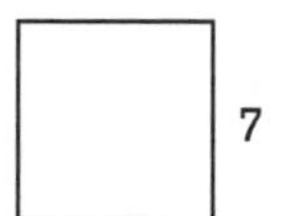

a. It is doubled.

b. It is quadrupled.

c. It has an eightfold increase.

d. It is one quarter of the original area.

32. Which proportion represents the percent of a week that four days is equivalent to?

a. $\frac{7}{4} = \frac{x}{100}$

b. $\frac{4}{7} = \frac{x}{100}$

c. $\frac{4}{7} = \frac{x}{30}$

d. $\frac{7}{30} = \frac{x}{100}$

33. If the volume in a water tank, V, is increased by 25%, which of the following expressions represents the new volume of water?

a. $V + \frac{1}{4}V$

b. 1.25 V

c. V + .25V

d. all of the above

34. The cylindrical can below has a label that wraps around the entire can with no overlap. Calculate the area of the label using $\pi = \frac{22}{7}$.

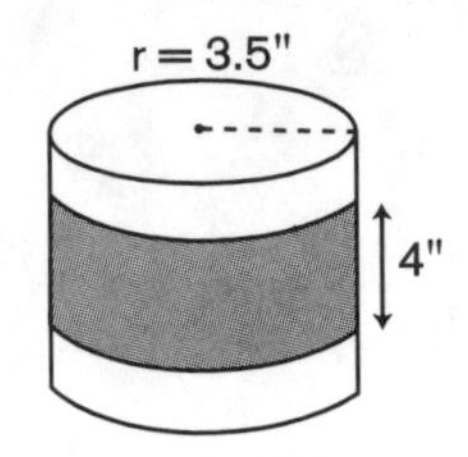

a. 14 in^2
b. 22 in^2
c. 88 in^2
d. 100 in^2

35. In the diagram below, what is the area of the deck (shown in gray)?

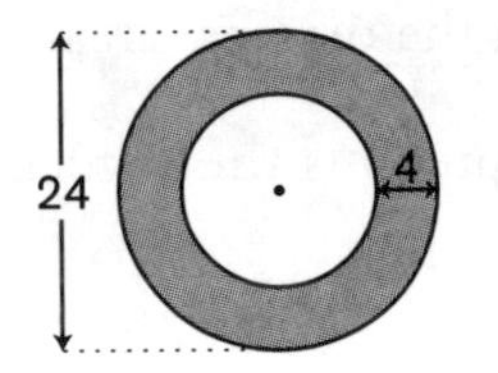

a. 144π ft^2
b. 108π ft^2
c. 92π ft^2
d. 80π ft^2

answers

1. d. Here you use the formula $w \times d = W \times D$. Substituting in the given values, you have $6 \cdot 8 = 12 \cdot x$, or $48 = 12x$. Divide both sides by 12 to yield $x = 4$ ft, choice **d**.

2. c. He will save 25% off the original \$80. $25\% = \frac{1}{4}$, so he saves $\frac{1}{4}$ of \$80. The following diagram represents this situation:

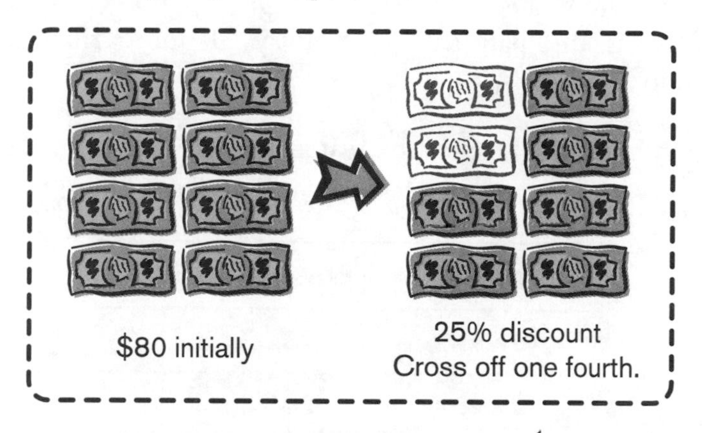

Thus, the discounted jacket would cost $\frac{1}{4} \cdot \$80 = \$60$. Alternatively, you can figure that Ed will pay 75% of the original price, or $\frac{3}{4}$ of the original price. $\frac{3}{4} \cdot \$80 = \$60$.

3. b. Use the formula $\frac{change}{initial} = \frac{?}{100}$.

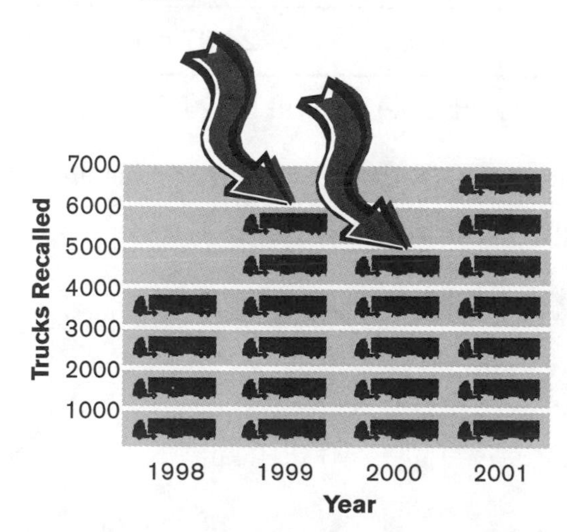

The change is 1,000 (note that the graph goes up to 6,000 in 1999, and only 5,000 in 2000). The initial equals 6,000. You get:

$$\frac{1000}{6000} = \frac{?}{100}$$

You can reduce the ratio on the left to $\frac{1}{6}$, so you have:

$$\frac{1}{6} = \frac{?}{100}$$

Cross multiplying, you have: $100 \bullet 1 = 6 \bullet ?$, or $100 = 6 \bullet ?$. Dividing both sides by 6, you get $? = 16\frac{2}{3}$. Thus, there was a $16\frac{2}{3}\%$ decrease.

4. **c.** When x is 0, y is 0. When $x = 2$, $y = 5$. And when $x = 4$, $y = 17$. What is the pattern? y is always 1 more than x^2. Thus, **c**, $y = x^2 + 1$ is correct.

5. **c.** Each hour, the level of the water increases by .21 cm. Let's add data for 8 A.M., 9 A.M., and 10 A.M. in the chart below:

Time	Water Height
5 A.M.	23.02 cm
6 A.M.	23.23 cm
7 A.M.	23.44 cm
8 A.M.	23.65 cm
9 A.M.	23.86 cm
10 A.M.	24.07 cm

Thus, at 10 A.M., the height will be 24.07 cm.

6. **b.** If you count up the number of squares that are shaded, you will find that $\frac{12}{16}$ are shaded. $\frac{12}{16} = \frac{3}{4}$.

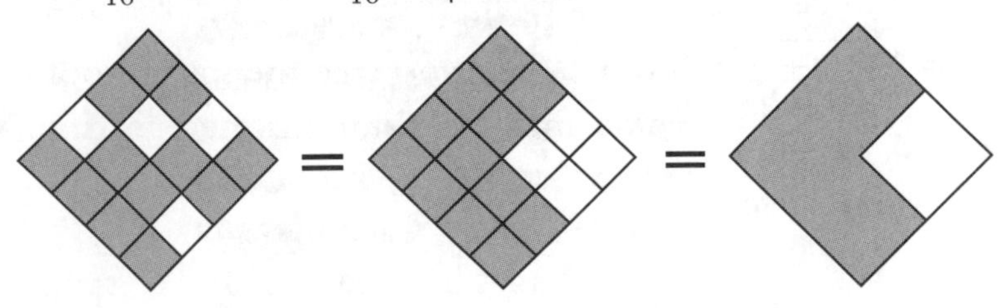

7. **b.** Triangle ABC and triangle DAE are similar. This means that their sides will be in proportion. Side AB will be in proportion with side

AD. On the figure we can see that AB = 3. We are given that DB = 6, so we know that AD = 9. Thus the triangles are in a 3:9 ratio, which reduces to a 1:3 ratio. This helps us because if AC = 2, then AE will be three times as long, or 6.

8. a. It is easy to solve this kind of question if you draw a diagram:

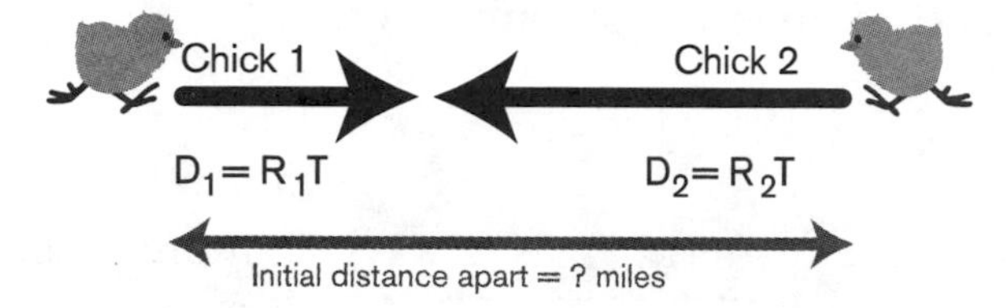

You know that the total distance (? miles) is equal to $D_1 + D_2$.

$$? = D_1 + D_2$$

You know $D_1 = R_1 \times T$ and $D_2 = R_2 \times T$, so you can rewrite the above equation as:

$$? = (R_1 \times T) + (R_2 \times T)$$

You can put in 2 for T, and the rates of each chick (3 mph & 4 mph), to yield:

$$? = 3(2) + 4(2)$$

$$? = 6 + 8$$

$$? = 14 \text{ miles}$$

Thus, choice **a** is correct.

9. d. To find the reciprocal of any given fraction, just switch the numerator and denominator: $\frac{1}{5}$ reciprocal ➔ $\frac{5}{1} = 5 \div 1 = 5$.

10. d. $x = \sqrt{10 + 1} = \sqrt{11}$. When you take the square root of 11, you get a number with a decimal extension that never terminates or repeats. This means that the x cannot be represented as a ratio of two integers, and is thus an irrational number.

11. c. 3.1×3^{10} is 9 times as large as 3.1×3^8. This is because when you compare the exponents, 3^{10} would be ten threes multiplied together, and 3^8 would be eight threes multiplied together. Thus, the first is 3×3, or 9 times as large.

12. **b.** The first thing you want to do is isolate your variable. This means you want to combine your x terms on one side of the equation, and your numbers on the other side of the equation. Below, add $8x$ to both sides in order to combine x terms:

$$\begin{array}{rl} -8x + 2 = & 5x + 1 \\ +\ 8x \quad & +\ 8x \\ \hline 2 = & 13x + 1 \end{array}$$

Now you will subtract 1 from both sides in order to isolate the x term.

$$\begin{array}{rl} 2 = & 13x + 1 \\ -\ 1 \quad & -1 \\ \hline 1 = & 13x \end{array}$$

Finally, divide both sides by 13 to get $x = \frac{1}{13}$, or choice **b**.

13. **c.** Use the chart to note that Overseas brought in 9% of the total:

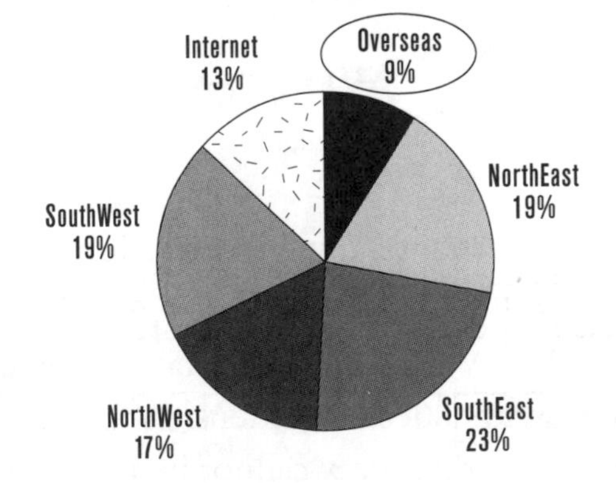

You know the total is \$270,000, so you take 9% of this amount:

$$.09 \times \$270{,}000 = \$24{,}300$$

14. **a.** First calculate the number of possible outcomes. The possibilities are: 1, 2, 3 ,4, 5, or 6. Thus, there are 6 possible outcomes. Next, underline the outcomes that satisfy the condition given in the question. Which of these numbers are multiples of 3? 1, 2, **3**, 4, 5, **6**. Notice that only two outcomes out of a total of six outcomes are multiples of 3. Thus the answer would be $\frac{2}{6}$, which reduces to $\frac{1}{3}$.

15. b. Here, the order matters, and you have four available places: ${}_4P_4 = 4 \times 3 \times 2 \times 1 = 24$. Alternatively, if you listed the possibilities, you would get:

A B Sammy C D	A D Sammy B C
A B Sammy D C	A D Sammy C B
B A Sammy C D	D A Sammy B C
B A Sammy D C	D A Sammy C B
A C Sammy B D	B D Sammy A C
A C Sammy D B	B D Sammy C A
C A Sammy B D	D B Sammy A C
C A Sammy D B	D B Sammy C A
C B Sammy A D	C D Sammy A B
C B Sammy D A	C D Sammy B A
B C Sammy A D	D C Sammy A B
B C Sammy D A	D C Sammy B A

16. a. The diamond will tessellate:

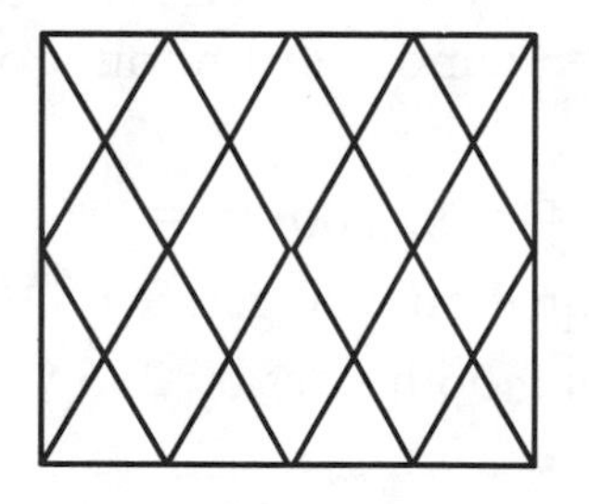

The other choices listed do NOT tessellate because there are gaps between the shapes as you tile them atop a surface:

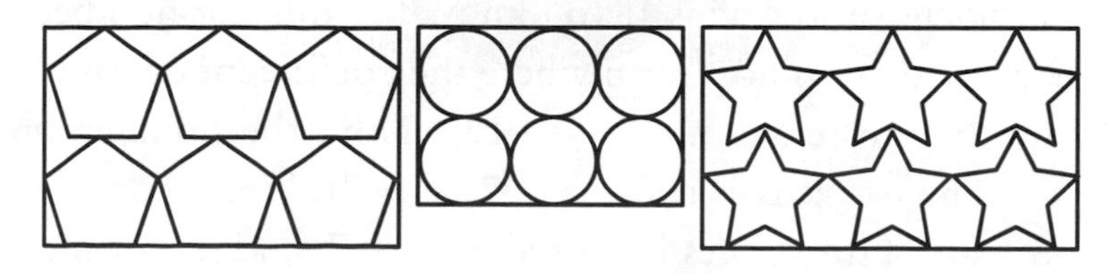

17. d. "Five plus a number" can be expressed as $n + 5$. You multiply $(n + 5)$ by "three less than the number," or $n - 3$. Thus, you have $(n + 5)(n - 3)$. Using FOIL, you get:

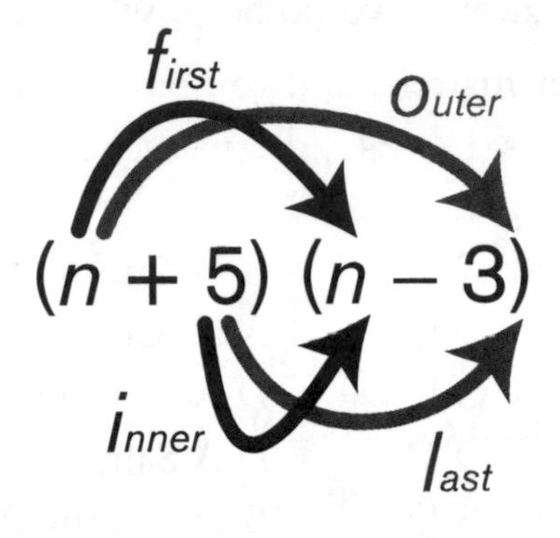

$= n^2 - 3n + 5n - 15$. This simplifies to $n^2 + 2n - 15$.

18. b. Use the trigonometric ratio $tan = \frac{\text{opposite}}{\text{adjacent}}$. The tangent of 60° equals $\sqrt{3}$ (you can use your calculator). Substituting the given values into the ratio, you get: $\sqrt{3} = \frac{h}{2\sqrt{3}}$. You cross-multiply to get $h = \sqrt{3} \times 2\sqrt{3} = 2\,(\sqrt{3})^2 = 2 \cdot 3 = 6$ meters.

19. c. The formula I = PRT means:

Interest = *principal* × *rate of interest* × *time*

Where *principal* = your original amount of money (in dollars), and *time* is in years.

In this question, P = \$3,200, R = 8%, or .08, and I = \$768. Substituting into the equation, you get: \$768 = \$3,200 × .08 × T, or \$768 = 256T, and dividing both sides by 256 yields T = 3. Thus, the time = 3 years.

20. a. First, let's take a look at the top part of the expression: $x^2 - 49$ can be factored into two sets of parentheses: $(x \pm ?)(x \pm ?)$. Because the coefficient on the x^2 is 1, you know the missing numbers add to 0 (because there is no x term where the coefficient of x is 0) and multiply to 49 (the *lone number* is –49). Thus, the numbers are –7 and + 7. The top part is really $(x - 7)(x + 7)$. Next, consider the bottom part of the expression, or $x^2 + 6x - 7$, and set up another set of parentheses: $(x \pm ?)(x \pm ?)$. Here the missing numbers add to 6 (the coefficient on the x term is 6) and multiply to –7 (the *lone number* is −7). Thus, the numbers are −1 and +7. The bottom part is really $(x - 1)(x + 7)$. So you put the top over the bottom to get:

$$\frac{(x - 7)(x + 7)}{(x - 1)(x + 7)}$$

Finally, notice that you can cross out an $(x + 7)$ on top and an $(x + 7)$ on the bottom. You are left with:

$$\frac{(x - 7)}{(x - 1)}$$

21. d. Set up a permutation to solve this question. Here, the order matters, and you have eight available places.

22. b. The line that crosses both parallel lines will create the same angles for both lines. There is an angle marked "$x - 6$" under line segment $\overline{AB}$, so we can mark an angle "$x - 6$" under line segment $\overline{CD}$. Now, notice that $2x$ and $x - 6$ combine to make a straight line. Since a straight line is 180 degrees, we can write: $2x + (x - 6) = 180$, or $3x - 6 = 180$, or $3x = 186$, or $x = 62°$, answer choice **b.**

23. a. The formula for calculating the mean (average) is:

$$\text{Mean} = \frac{\text{sum of all values}}{\text{\# of values}}$$

The sum of all the values given is: $75 + 81 + 93 + 77 + 84 = 410$. The number of values (scores) is 5. Thus, the mean $= \frac{410}{5} = 82$.

24. d. The perimeter formula for a rectangle is $P = 2l + 2w$. Here the length is $2 - a$, and the width is a. Putting these values into our formula we get $L = 2(2 - a) + 2(a) = 4 - 4a + 4a = 4$.

25. c. When you substitute $a = 2$, $b = -1$, and $c = \frac{1}{2}$, into the expression

$$\frac{2a - b + 5}{c}$$

you get:

$$\frac{2(2) - (-1) + 5}{\frac{1}{2}}$$

which simplifies to:

$$\frac{4 + 1 + 5}{\frac{1}{2}}$$

which equals:

$$\frac{10}{\frac{1}{2}}$$

10 divided by $\frac{1}{2}$ is the same as 10 times $\frac{2}{1}$, or 10×2, which equals 20.

26. b. Just remember that when taking a fraction **of** some number, you are actually **multiplying** that number by the fraction. Let's look at what happens hour by hour:

Time	# candies
Start	504
8–9 o'clock	minus $\frac{1}{8} \cdot 504$, or $-63 = 441$
9–10 o'clock	minus $\frac{2}{9} \cdot 441$, or $-98 = 343$
10–11 o'clock	minus $\frac{1}{7} \cdot 343$, or $-49 = 294$

27. b. Percents are always *out of 100.* The chart even reminds you that the total is 100%. This means that $18\% + 10\% + 65\% + x\% = 100\%$. To solve, just subtract all the percents on the right of the equation from 100. You get: $x = 100 - 18 - 10 - 65 = 7$. Thus, $x = 7$.

28. c. Because the term *mean* is another way to say *average*, you know that **d** must be wrong. You will need to calculate the *mean* and the *median*. The formula for calculating the mean (average) is:

$$\text{Mean} = \frac{\text{sum of all values}}{\text{\# of values}}$$

Here the sum of all the values is $12 + 14 + 15 + 19 + 20 + 22 = 102$, and the number of values is 6. Thus the average is $\frac{102}{6} = 17$. To find the median, you circle the two middle numbers and divide by 2. Thus, $\frac{15 + 19}{2} = 17$. Therefore, the mean equals the median.

29. b. To figure out the probability for the given outcome, you need to calculate the total possible outcomes. You know that the record company brought 300 hip-hop CDs, 500 alternative rock CDs, 200 easy listening CDs, and 400 country CDs. The total possible outcomes equal 300 + 500 + 200 + 400 = 1,400. The outcomes that fit the criteria in the question = 200. This is because 200 easy listening CDs will be given out. This means that the chance of getting an easy listening CD will be $\frac{200}{1{,}400}$. This reduces to $\frac{1}{7}$.

30. c. Given $\log_{10} 10^5$, you spiral through and say:

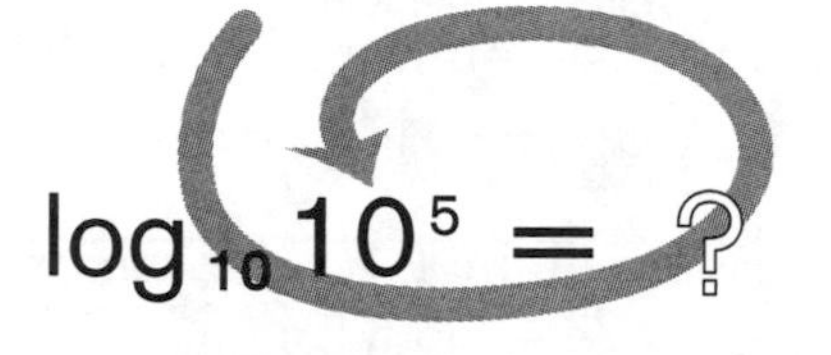

"10 to what power is 10^5?"

The power would be 5, so $\log_{10} 10^5 = 5$.

31. b. The area of the square is $s^2 = 7^2 = 49$. If you double the side, the new area equals $s^2 = 14^2 = 196$. 49 times 4 equals 196, so you quadrupled the area. Notice that you can "see" this effect below:

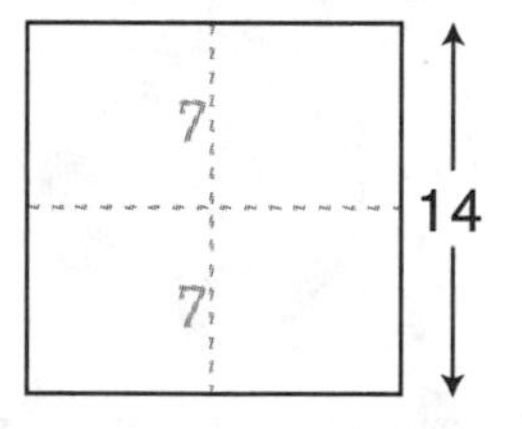

Algebraically, any square that has its side doubled will also have its area quadrupled because when you compare s^2 to $(2s)^2$ you get:

$$s^2 \text{ versus } (2s)^2$$

or

$$s^2 \text{ versus } 4s^2$$

$4s^2$ is obviously 4 times the s^2.

32. b. You set up the proportion as follows:

$$\frac{4 \text{ days}}{7 \text{ days per week}} = \frac{x}{100}$$

Thus, choice **b** is correct.

33. d. You need to take V and add 25% of V. 25% = .25 or $\frac{25}{100} = \frac{1}{4}$. Thus, choice **a**, V + $\frac{1}{4}$ V is true. This is the same as V + .25 V, which is choice **c.** If you actually add choice **c**, you get choice **b.**

34. c. Note the dimensions of the label (when peeled from the can):

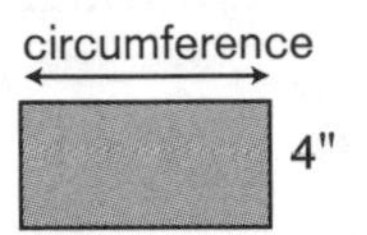

The area of the label will be L × W = circumference × 4 in. C = $2\pi r = 2 \bullet \frac{22}{7} \bullet 3.5 = 7 \bullet \frac{22}{7} = 22$ in. Thus, the area = 22 in × 4 in = 88 in^2.

35. d. You need to find the area of the big (outer) circle and subtract the area of the small (inner) circle. You draw in the radius of the "big" circle:

r big = 12

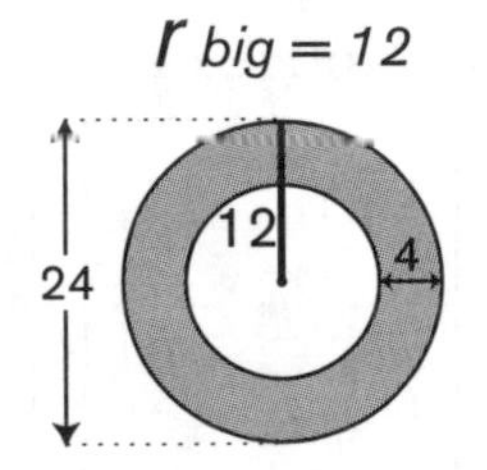

You can also draw in the radius of the small circle:

r small = 12 – 4 = 8

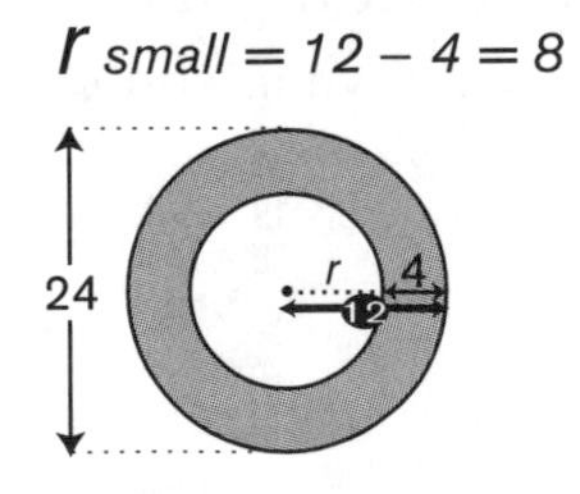

The big circle has an area equal to $A = \pi r^2 = \pi(12)^2 = 144\pi$. The small circle has an area of $A = \pi r^2 = \pi(8)^2 = 144\pi$. Thus, the area of the shaded region is $144\pi - 64\pi = 80\pi$ ft^2.